Introduction

The aim of the *Primary Mathematics* curriculum is to allow students to develop their ability in mathematical problem solving. This includes using and applying mathematics in practical, real-life situations as well as within the discipline of mathematics itself. Therefore the curriculum covers a wide range of situations from routine problems to problems in unfamiliar contexts to open-ended investigations that make use of relevant mathematical concepts.

An important feature of learning mathematics with this curriculum is the use of a concrete introduction to the concept, followed by a pictorial representation, followed by the abstract symbols. The textbook does supply some concrete introductory situations, but you, as the teacher, should supply a more concrete introduction when applicable. The textbook then supplies the pictorial and abstract aspects of this progression. For some students a concrete illustration is more important than for other students.

This guide includes the following :

- **Scheme of Work**: A table with a suggested weekly schedule, the primary objective for each lesson, and corresponding pages from the textbook, workbook, and guide.
- **Manipulatives**: A list of manipulatives used in this guide.
- **Objectives**: A list of objectives for each chapter.
- **Vocabulary**: A list of new mathematical terms for each chapter.
- **Notes**: An explanation of what students learned in earlier levels, the concepts that will be covered in the chapter, and how these concepts fit in with the program as a whole.
- **Material**: A list of suggested manipulatives that can be used in presenting the concepts in each chapter.
- **Activity**: Teaching activities to introduce a concept concretely or to follow up on a concept in order to clarify or extend it so that students will be more successful with independent practice.
- **Discussion**: The opening pages of the chapter and tasks in the textbook that should be discussed with the student. A scripted discussion is not provided. You should follow the material in the textbook. Additional pertinent points that should be included in the discussion are given in this guide.
- **Practice**: Tasks in the textbooks students can do as guided practice or as an assessment to see if they understood the concepts covered in the teaching activity or the discussions.
- **Workbook**: Workbook exercise that should be done after the lesson.
- **Reinforcement**: Additional activities that can be used if your student needs more practice or reinforcement of the concepts. This includes references to the exercises in the optional *Primary Mathematics Extra Practice* book.
- **Games**: Optional simple games that can be used to practice skills.

- **Enrichment**: Optional activities that can be used to further explore the concepts or to provide some extra challenge.
- **Tests:** References to the appropriate tests in the *Primary Mathematics Tests* book.
- **Answers:** Answers to all the textbook tasks and workbook problems, and many fully worked solutions. Answers to textbook tasks are provided within the lesson. Answers to workbook exercises for the chapter are located at the end the chapters in the guide.
- **Mental Math**: Problems for more practice with mental math strategies.
- **Appendix**: Pages containing drawings and charts that can be copied and used in the lessons.

The textbook and workbook both contain a review for every unit. You can use these in any way beneficial to your student. For students who benefit from a more continuous review, you can assign three problems or so a day from one of the practices or reviews. Or, you can use the reviews to assess any misunderstanding before administering a test. The reviews, particularly in the textbook, do sometimes carry the concepts a little farther. They are cumulative, and so allow you to refresh your student's memory or understanding on a topic that was covered earlier in the year.

In addition, there are supplemental books for *Extra Practice* and *Tests*. In the test book, there are two tests for each section. The second test is multiple choice. There is also a set of two cumulative tests at the end of each unit. You do not need to use both tests. If you use only one test, you can save the other for review or practice later on. You can even use the review in the workbook as a test and not get the test book at all. So there are plenty of choices for assessment, review, and practice.

The mental math exercises that go along with a particular chapter or lesson are listed as reinforcement in the lesson. They can be used in a variety of ways. You do not need to use all the mental math exercises listed for a lesson on the day of the lesson. In some review lessons, they can be used for the independent work, since there is not always a workbook exercise. You can have your student do one mental math exercise a day, repeating some of them, at the start of the lesson or as part of the independent work. You can do them orally, or have your student fill in the blanks. You can have your student do a 1-minute "sprint" at the start of each lesson using one mental math exercise for several days to see if he or she can get more of the problems done each successive day. You can use the mental math exercises as a guide for creating additional "drill" exercises.

The "Scheme of Work" on the next few pages is a suggested weekly schedule to help you keep on track for finishing the textbook in about 18 weeks. No one schedule or curriculum can meet the needs of all students equally. For some chapters, your student may be able to do the work more quickly, and for others more slowly. Take the time your student needs on each topic and each lesson. For students with a good mathematical background, each lesson in this guide will probably take a day. For others, some lessons which include a review of previously covered concepts may take more than a day. There are a few lessons that are only review that your student may not need.

Use the reinforcement or enrichment activities at your discretion and according to your student's needs.

This printing of this guide was written when the latest printing of the textbook, workbook, Extra Practice, and Tests were in 2008. New printings of these books may have corrected errors or may have slight changes, which will be incorporated into later printings of the guide.

Scheme of Work

Textbook: *Primary Mathematics Textbook* 4A, Standards Edition
Workbook: *Primary Mathematics Workbook* 4A, Standards Edition
Guide: *Primary Mathematics Home Instructor's Guide* 4A, Standards Edition (this book)
Extra Practice: *Primary Mathematics Extra Practice* 4, Standards Edition
Tests: *Primary Mathematics Tests* 4A, Standards Edition

Week		Objectives	Text book	Work book	Guide
Unit 1: Whole Numbers					
	Chapter 1: Ten Thousands, Hundred Thousands and Millions				1-6
1	1	♦ Understand place value for numbers within 1 million. ♦ Write numbers of up to six digits in standard and expanded form. ♦ Read and write 5-digit and 6-digit numbers and corresponding number words.	8-13	7-9	7-8
	2	♦ Relate each digit in a 5-digit or 6-digit number to its place value.	13-15	10-12	9
	3	♦ Understand place value for numbers within 1 billion. ♦ Write numbers of up to nine digits in standard and expanded form. ♦ Read and write numbers within 1 billion and corresponding number words.	15-17	13-14	10-11
	4	♦ Count on or back by multiples of ten. ♦ Recognize and extend number patterns involving counting on or back by multiples of ten. ♦ Locate 4-digit and 5-digit numbers on a scaled number line. ♦ Compare and order numbers within 1 billion.	17-19	15-16	12-13
		Extra Practice, Unit 1, Exercise 1, pp. 7-10			
	5	♦ Review some mental math strategies for adding and subtracting numbers within 100.			14-15
2	6	♦ Use place-value concepts in mental computation. ♦ Use a letter to stand for an unknown in a simple equation and find the value of the letter.	19		16-17
	7	♦ Review finding a missing part in equations. ♦ Find the missing part in equations that involve mental math and recognizing place value.	19	17-18	18-19
	8	♦ Practice.	20-21		20
		Tests, Unit 1, 1A and 1B, pp. 1-6			
	Answers to Workbook Exercises 1-5				21-22

Week		Objectives	Text book	Work book	Guide
	Chapter 2: Approximation				23
	1	♦ Round numbers within 1 billion to the nearest ten, hundred, thousand, ten thousand, hundred thousand, and billion.	22-24	19-20	24
	2	♦ Practice.	25		25
		Extra Practice, Unit 1, Exercise 2, pp. 11-12			
		Tests, Unit 1, 2A and 2B, pp. 7-10			
	Chapter 3: Factors				26-27
3	1	♦ Use rectangular arrays to understand the meaning of factors. ♦ Understand the meaning of prime and composite numbers. ♦ Use arrays or multiplication facts to find factors of numbers within 100.	26-28	21-22	28
	2	♦ Use division to determine if a number is a factor of another number. ♦ Find the common factors of two numbers within 100. ♦ Find a missing factor in equations with two factors.	29-30	23-24	29
	3	♦ Understand that whole numbers can break down into factors in different ways. ♦ Find the missing factor in equations with more than two factors.	31	25	30
		Extra Practice, Unit 1, Exercise 3, pp. 13-14			
		Tests, Unit 1, 3A and 3B, pp. 11-14			
	Chapter 4: Multiples				31-32
	1	♦ Understand the meaning of multiples. ♦ Relate multiples to factors. ♦ Determine if a number is a multiple of another. ♦ List multiples of numbers within 10. ♦ Learn divisibility rules for 2, 3, and 5. ♦ Find common multiples of two numbers.	32-35	26-27	33-34
		Extra Practice, Unit 1, Exercise 4, pp. 15-16			
	2	♦ Practice. ♦ Learn divisibility rules for 6 and 9.	36-37		35
		Tests, Unit 1, 4A and 4B, pp. 15-17			
	Answers to Workbook Exercises 6-9				36

Week		Objectives	Text book	Work book	Guide
	Chapter 5: Order of Operations				37
4	1	♦ Solve expressions with more than two numbers that involve just one of the four operations. ♦ Solve expressions that involve addition and subtraction. ♦ Solve expressions that involve multiplication and division. ♦ Learn order of operations. ♦ Solve expressions that involve all four operations.	38-39	28-29	38-39
	2	♦ Solve expressions with parentheses.	39-40	30-31	40
	3	♦ Write expressions for two-step word problems that show both steps in a single expression. ♦ Understand that performing the same operation on both sides of an equation does not change the equality.	40-41	32-33	41
		Extra Practice, Unit 1, Exercise 5, pp. 17-18			
		Tests, Unit 1, 5A and 5B, pp. 19-22			
	Chapter 6: Negative numbers				42
	1	♦ Use negative numbers in practical situations. ♦ Locate negative numbers on a number line.	42-44	34-35	43
	2	♦ Compare integers. ♦ Put integers in order.	44-45	36-37	44
		Extra Practice, Unit 1, Exercise 6, pp. 19-20			
5	3	♦ Practice.	46-47		45
		Tests, Unit 1, 6A and 6B, pp. 23-26			
	Review 1		48-50	38-39	46
		Tests, Unit 1 Cumulative Tests A and B, pp. 27-32			
	Answers to Workbook Exercises 10-14 and Review 1				47-48
Unit 2: The Four Operations of Whole Numbers					
	Chapter 1: Addition and Subtraction				49-51
	1	♦ Add multi-digit numbers using the standard algorithm. ♦ Subtract multi-digit numbers using the standard algorithm.	51-52	40-41	52
	2	♦ Review finding the missing numbers in equations involving addition or subtraction. ♦ Use mental computation to make 1000.	53-54	42-43	53-54

Week		Objectives	Text book	Work book	Guide
6	3	♦ Use mental math strategies to subtract from 10, 100, and 1000. ♦ Use mental math strategies to add and subtract some multi-digit numbers.	54-55	44-46	55-56
	4	♦ Use approximation to estimate the answer to addition and subtraction problems. ♦ Determine if a problem requires an estimated or an exact answer.	55-56	47-48	57
	5	♦ Review modeling method for solving word problems involving addition and subtraction. ♦ Solve word problems involving addition and subtraction.	57-58	49-50	58-59
		Extra Practice, Unit 2, Exercise 1, pp. 23-24			
		Tests, Unit 2, 1A and 1B, pp. 33-36			
	Answers to Workbook Exercises 1-5				60-61
	Chapter 2: Multiplication and Division				62-68
	1	♦ Review multiplication of a number within 10,000 by a 1-digit number. ♦ Review modeling methods for solving word problems involving multiplication. ♦ Use approximation to estimate the product.	59, 61	51	69
	2	♦ Review division of a number within 10,000 by a 1-digit number. ♦ Review modeling methods for solving word problems involving division. ♦ Divide by 10. ♦ Use approximation to estimate the quotient.	60, 62-64	52-53	70-71
7	3	♦ Solve word problems.	64-66	54-55	72
		Extra Practice, Unit 2, Exercise 2, pp. 25-26			
	4	♦ Practice.	67		73
		Tests, Unit 2, 2A and 2B, pp. 37-42			
	Chapter 3: Multiplication by a 2-Digit Number				74
	1	♦ Multiply a multi-digit number by tens. ♦ Multiply a multi-digit number by tens and ones.	68-70	56	75
	2	♦ Multiply a multi-digit number by a 2-digit number. ♦ Use various mental math strategies to multiply some numbers.	70-72	57	76-77

Week		Objectives	Text book	Work book	Guide
	3	♦ Estimate the product of multiplication.	72	58-62	78
		Extra Practice, Unit 2, Exercise 3, pp. 27-28			
8	4	♦ Practice.	73		79
		Tests, Unit 2, 3A and 3B, pp. 43-46			
	Review 2		74-76	63-66	80
		Tests, Units 1-2 Cumulative Tests A and B, pp. 47-53			
	Answers to Workbook Exercises 6-12 and Review 2				81-82
Unit 3 : Fractions					
	Chapter 1: Equivalent Fractions				83-84
	1	♦ Review finding equivalent fractions. ♦ Review finding the simplest form of a fraction.	77-78	67-68	85-86
	2	♦ Review comparing and ordering fractions.	79	69-70	87-88
		Extra Practice, Unit 3, Exercise 1, pp. 35-36			
9	3	♦ Practice.	80		89
		Tests, Unit 3, 1A and 1B, pp. 55-59			
	Answers to Workbook Exercises 1-2				90
	Chapter 2: Adding and Subtracting Fractions				91
	1	♦ Add like or related fractions.	81-84	71-72	92
	2	♦ Subtract like or related fractions.	85-86	73-74	93
	3	♦ Solve simple word problems involving addition and subtraction of fractions.	87	75-76	94-95
		Extra Practice, Unit 3, Exercise 2, pp. 37-42			
		Tests, Unit 3, 2A and 2B, pp. 61-65			
	Answers to Workbook Exercises 3-5				96
	Chapter 3: Mixed Numbers				97
	1	♦ Interpret mixed numbers as the sum of a whole number and a proper fraction. ♦ Read and interpret number lines with mixed numbers. ♦ Add a proper fraction to a whole number. ♦ Subtract a proper fraction from a whole number.	88-89	77-78	98
		Extra Practice, Unit 3, Exercise 3, pp. 43-44			
		Tests, Unit 3, 3A and 3B, pp. 67-71			

Week		Objectives	Text book	Work book	Guide
	Chapter 4: Improper Fractions				99
10	1	♦ Interpret improper fractions as repeated addition of unit fractions. ♦ Identify improper fractions.	90-91	79-80	100
	2	♦ Convert an improper fraction to a mixed number.	92	81-82	101
	3	♦ Convert a mixed number to an improper fraction.	92-93	83-85	102
		Extra Practice, Unit 3, Exercise 4, pp. 45-46			
	4	♦ Solve problems involving mixed numbers and improper fractions.	93	86-87	103
		Tests, Unit 3, 4A and 4B, pp. 73-77			
	Chapter 5: Fractions and Division				104
	1	♦ Relate fractions to division.	94-96	88-89	105
		Extra Practice, Unit 3, Exercise 5, pp. 47-48			
11	2	♦ Practice.	97		106
		Tests, Unit 3, 5A and 5B, pp. 79-82			
	Answers to Workbook Exercises 6-11				107-108
	Chapter 6: Fraction of a Set				109
	1	♦ Review fraction of a set.	98-99	90-94	110-111
	2	♦ Find the fraction of a set.	100	95-97	112
	3	♦ Find the fractional part of a whole.	101	98-99	113
	4	♦ Solve word problems.	102	100-101	114
12	5	♦ Solve word problems.	102-103	102-104	115
	6	♦ Solve word problems.	104	105-106	116
	7	♦ Solve word problems.	104-105	107-109	117
		Extra Practice, Unit 3, Exercise 6, pp. 49-52			
		Tests, Unit 3, 6A and 6B, pp. 83-87			
	Review 3		106-109	110-116	118
		Tests, Units 1-3 Cumulative Tests A and B, pp. 89-95			
	Answers to Workbook Exercises 12-18 and Review 3				119-121

Week		Objectives	Text book	Work book	Guide
Unit 4: Geometry					
	Chapter 1: Right Angles **Chapter 2: Measuring Angles**				122
13	1	♦ Relate quarter turn, half turn, three-quarter turn and whole turn to right angles. ♦ Classify angles as right, acute, or obtuse.	110-111	117-120	123
		Extra Practice, Unit 4, Exercise 1, pp. 61-62			
		Tests, Unit 4, 1A and 1B, pp. 97-101			
	2	♦ Measure angles smaller than 180°.	112-113	121-122	124
	3	♦ Estimate angles. ♦ Draw angles smaller than 180°.	113	123-127	125
	4	♦ Measure and draw angles greater than 180°.	114-115	128-131	126
		Extra Practice, Unit 4, Exercise 2, pp. 63-68			
		Tests, Unit 4, 2A and 2B, pp. 103-109			
	Chapters 3: Perpendicular Lines **Chapter 4: Parallel Lines**				127
	1	♦ Identify perpendicular lines.	116-117	132-133	128
14	2	♦ Construct perpendicular lines.	118	134-135	129
		Extra Practice, Unit 4, Exercise 3, pp. 69-70			
		Tests, Unit 4, 3A and 3B, pp. 111-116			
	3	♦ Identify parallel lines.	119-120	136-137	130
	4	♦ Construct parallel lines.	121	138-139	131
		Extra Practice, Unit 4, Exercise 4, pp. 71-72			
		Tests, Unit 4, 4A and 4B, pp. 117-124			
	Answers to Workbook Exercises 1-8				132
	Chapter 5: Quadrilaterals				133-134
	1	♦ Identify common polygons, parallelograms, and trapezoids. ♦ Recognize rectangles, squares, and rhombuses as types of parallelograms. ♦ Find the unknown lengths of sides of parallelograms given the length of other sides.	122-124	140-141	135
		Extra Practice, Unit 4, Exercise 5, pp. 73-74			
		Tests, Unit 4, 5A and 5B, pp. 125-131			

Week		Objectives	Text book	Work book	Guide
	Chapter 6: Triangles				136
15	1	♦ Identify equilateral, isosceles, and scalene triangles. ♦ Identify right, acute, and obtuse triangles. ♦ Find the unknown lengths of sides of triangles, given the lengths of other sides.	125-127	142-143	137
		Extra Practice, Unit 4, Exercise 6, pp. 75-76			
		Tests, Unit 4, 6A and 6B, pp. 133-138			
	Chapter 7: Circles				138
	1	♦ Identify the center, diameter, and radius of a circle. ♦ Measure the radius or diameter of a circle. ♦ Find the radius of a circle given its diameter. ♦ Find the diameter of a circle given its radius.	128-129	144-145	139
		Extra Practice, Unit 4, Exercise 7, pp. 77-78			
		Tests, Unit 4, 7A and 7B, pp. 139-142			
	Chapter 8: Solid Figures				140
	1	♦ Visualize cubes, prisms, pyramids, and cylinders from two-dimensional drawings. ♦ Determine the number and shapes of the faces of a solid from a two-dimensional drawing.	130-131	146-147	141
		Extra Practice, Unit 4, Exercise 8, pp. 79-80			
		Tests, Unit 4, 8A and 8B, pp. 143-148			
	Chapter 9: Nets				142
	1	♦ Form solids from nets. ♦ Identify nets of cubes.	132=134	148-149	143
	2	♦ Identify the net formed from a solid. ♦ Identify the solid represented by a net.	135-136	150-155	144
		Extra Practice, Unit 4, Exercise 9, pp. 81-82			
		Tests, Unit 4, 9A and 9B, pp. 149-156			
	Answers to Workbook Exercises 9-15				145
16	**Review 4**		137-140	156-161	146
		Tests, Units 1-4, Cumulative A and B, pp. 157-167			
	Answers to Workbook Review 4				147

Materials

It is important to introduce the concepts concretely, but it is not important exactly what manipulatives are used. A few possible manipulatives are suggested here. The linking cubes and playing cards will be used in many levels of *Primary Mathematics*.

Whiteboard and Dry-Erase Markers
A whiteboard that can be held is useful in doing lessons while sitting at the table (or on the couch). Students can work problems given during the lessons on their own personal boards.

Multilink cubes
These are cubes that can be linked together on all 6 sides.

Base-10 set
A set usually has 100 unit-cubes, 10 or more ten-rods, 10 hundred-flats, and 1 thousand-block. These are used rarely at this level, so are optional.

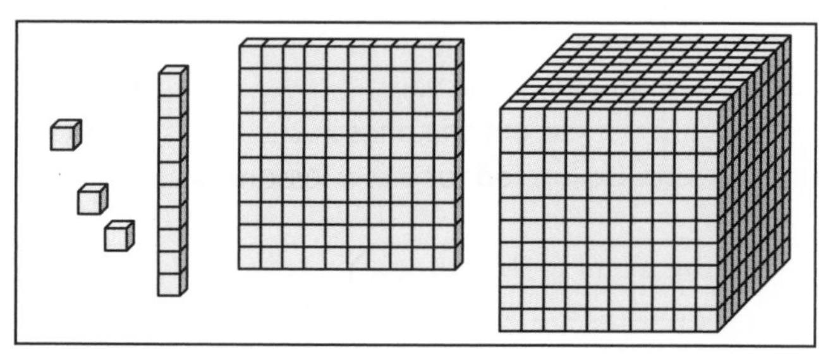

Round counters
To use as place-value discs, unless you buy commercial place-value discs, make paper ones, or just draw them.

Place-value discs
Round discs with 1, 10, 100, or 1000 written on them. You can label round counters using a permanent marker. You need 20 of each kind. You can label a few with 10,000, 100,000, and 1,000,000 if needed to illustrate place-value concepts for larger numbers.

Place-value chart
You can draw a simple one on paper or a white-board. It should be large enough to use with number discs or base-10 blocks. At this level, you will only need columns past 1000 in the first chapter, and you will add those columns as you discuss place value.

Hundred Thousands	Ten Thousands	Thousands	Hundreds	Tens	Ones
100000	10000 10000 10000 10000 10000	1000 1000	100 100 100	10 10 10 10 10 10	1 1 1 1 1 1 1 1

Hundred-chart
A number chart from 1-100. Laminated ones are nice, since you can use a dry-erase marker, or you can copy the one in the appendix.

Number cubes
Cubes that you can label and throw, like dice. You need two. You can use regular dice and label them with masking tape, or buy cubes and labels.

Number cards
Cards with 0, 1, 2, 3, 4, 5, 6, 7, 8, 9, or 10 written on them. You need four sets for games. You can use playing cards with face cards removed. Use one set of face cards for 0, whiting out the J, Q, or K and replacing it with a 0. You can white out the Ace and replace it with a 1. Or create a deck of number cards from index cards or blank playing cards.

Place-Value Cards
Place-value cards show the value of each digit and can be fitted on top of each other to form a numeral. Copy or cut out the ones in the appendix, or make your own with index cards, or buy them. They are only used for a few lessons at the beginning and are optional.

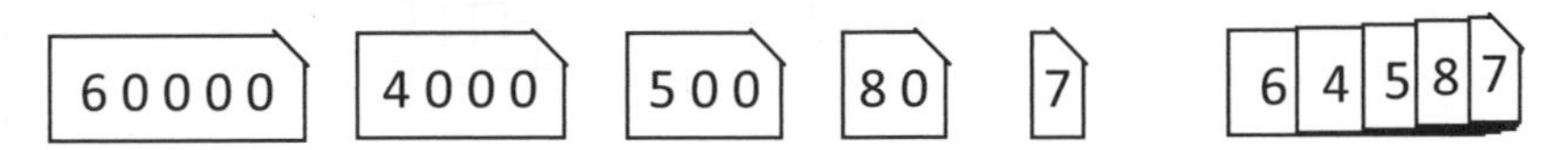

Graph paper
Centimeter graph paper. There is one you can copy in the appendix.

Geometry tools
Ruler
Set square (plastic triangle with a 90° angle)
Protractor
Compass (for drawing circles)

Supplements

The textbook and workbook provide the essence of the math curriculum. Some students profit by additional practice or more review. Other students profit by more challenging problems. There are several supplementary workbooks available at www.singaporemath.com.

If you feel it is important that your student have a lot of drill in math facts, there are many websites that generate worksheets according to your specifications, or provide on-line fact practice. Web sites come and go, but doing a search using the terms “math fact practice” will turn up many sites. Playing simple games is another way to practice math facts.

Unit 1 – Whole Numbers

Chapter 1 – Ten Thousands, Hundred Thousands and Millions

Objectives

- Understand place value for numbers within 1 billion.
- Relate each digit in a number of up to 9 digits to its place value.
- Write numbers of up to 9 digits in standard and expanded form and in words.
- Count on or back from a number within 1 billion by ones, tens, hundreds, thousands, ten thousands, hundred thousands, millions, ten millions, and hundred millions.
- Recognize and extend number patterns involving counting on or back by multiples of tens (tens, hundreds, thousands, ten thousands, hundred thousands, millions, ten millions, and hundred millions).
- Determine the scale of a number line and locate 4-digit and 5-digit numbers on a scaled number line.
- Compare and order numbers within 1 billion.
- Review some mental math strategies for adding and subtracting numbers within 100.
- Use place-value concepts in mental computation.
- Use a letter to stand for an unknown in a simple equation and find the value of the letter.
- Review finding a missing part in an equation.
- Find the missing number in equations that involve mental math and recognizing place value.

Vocabulary

- Standard form
- Expanded form
- Ten thousands
- Hundred thousands
- Millions
- Ten millions
- Hundred millions

Material

- Base-10 blocks, optional
- Place-value discs
- Place-value cards (appendix pp. a9-a14)
- Appendix p. a15
- Number cube or die
- Number cards, 0-9, 4 sets
- Mental Math 1-9

Notes

This chapter is primarily a review of concepts learned in earlier levels of *Primary Mathematics*, but extended to higher place values.

In *Primary Mathematics* 3A, students learned to relate 4-digit numbers to the place-value concept. In this chapter, the place-value concept is reinforced and extended to hundred thousands and then hundred millions; that is, to numbers of up to 9 digits.

Numbers are arranged in groups of three places called periods. The places within periods repeat (hundreds, tens, ones). The first period is called the ones, the second period the thousands, and the third period the millions. The number 123,456,789 is read as one hundred twenty-three million, four hundred fifty-six thousand, seven hundred eighty-nine. Commas are used to separate periods in both the standard form of the number and the number word. In *Primary Mathematics*, 4-digit numbers are written without the optional comma after the thousands, e.g. 4160.

The value of a digit is determined by its place in the number. This place-value concept is the basis of the base-10 system of numeration (the Hindu-Arabic system). We use ten digits (0 to 9) to write numbers, with each digit having a value that is ten times as much as the digit in the place to the right of it (and one tenth as much value as the same digit in the place to the left of it). The number 623,456 represents 6 hundred thousands, 2 ten thousands, 3 thousands, 4 hundreds, 5 tens, and 6 ones. The *place value* of the digit 3 is thousands, and its value is 3000.

Students who have used earlier levels of *Primary Mathematics* should not have too much difficulty extending place-value concepts to these larger numbers, and the pictures in the textbook should be sufficient. If your student has not used earlier levels, you may have to use concrete manipulatives such as place-value discs on a place-value chart or place-value cards more often during the lesson.

A place-value chart is a table divided into columns or adjacent places for ones, tens, hundreds, thousands, and so on. In earlier levels of *Primary Mathematics*, students used number discs to represent numbers. They learned that a disc with 10 on it was equivalent to 10 discs with 1 on them, a disc with 100 on it had the same value as ten discs with 10 on them, and so on. These number discs can be placed on the place-value chart to represent a number. Numerals can be used on the place-value chart as well, instead of discs, and discs can be drawn on the chart, instead of using actual discs.

Hundred Thousands	Ten Thousands	Thousands	Hundreds	Tens	Ones
100000	10000 10000 10000 10000 10000	1000 1000	100 100 100	10 10 10 10 10 10	1 1 1 1 1 1 1 1
1	5	2	3	6	8

The number shown on this chart is one hundred fifty-two thousand, three hundred sixty-eight, or 152,368. The digit 5 is in the ten-thousands place, and has a value of 50,000.

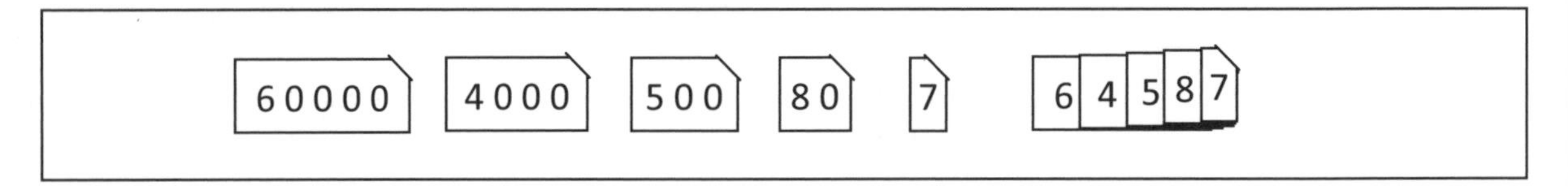

Place-value cards show the value of each digit and can be fitted on top of each other to form a numeral.

64,587 is called the **standard form** of a number. The **expanded form** is written as the sum of the values of each digit: 60,000 + 4000 + 500 + 80 + 7.

In this chapter, your student will also evaluate sequences of numbers where the numbers are increasing or decreasing in one of the places. In evaluating these sequences, she needs to focus on place value, and also understand in terms of place value what happens when counting on or back by that place value. For example, in the sequence 4,567,004; 4,568,004; 4,569,004 the digit in the thousands place increases by 1 each time. For the next number in the sequence, increasing the digit in the thousands place requires the digit in the ten-thousands place to change; 69 thousand becomes 70 thousand. The next number in the sequence is 4,570,004.

In *Primary Mathematics* 1B students learned to compare numbers within 100 and to use the symbol **>** for **greater than** and **<** for **less than**. In *Primary Mathematics* 2A they used the symbols with 3-digit numbers and in *Primary Mathematics* 3A they used them with 4-digit numbers. In this chapter, your student will compare larger numbers.

We can compare numbers by first comparing the digits in the highest place value. If they are the same, we then compare the digits in the next highest place value, and so on. Be sure to give your student the opportunity to compare numbers with a different number of digits.

↓ ↓ ↓ ↓
2, 5 4 5, 6 3 8
2, 5 4 6, 5 3 8
6 thousands > 5 thousands, so
2,546,538 > 2,545,638
2,545,638 < 2,546,538

In *Primary Mathematics* 3A students learned how to find the scale of a number line and locate 2-digit, 3-digit, and 4-digit numbers on number lines. In this chapter, your student will review finding the scale of the number line, and locate 5-digit numbers on a number line.

In this chapter your student will add, subtract, multiply, and divide large numbers with no more than 2 non-zero digits mentally in order to focus on place value. By now he should be able to add and subtract 2-digit numbers mentally and should know all the multiplication and division facts through 10 x 10. Some mental math strategies are listed below and will be briefly reviewed in the lessons, but if your student does not know how to add and subtract 2-digit numbers mentally and does not know the multiplication and division facts, you may want to consider having him do *Primary Mathematics* 3A.

When discussing strategies for mental math, you can draw number bonds to show how the numbers can be decomposed to make the calculations easier, but do not require your student to draw the bonds unless she is poor at mental math. The number bonds are meant to help her understand the strategies as she learns them, not to become another paper and pencil algorithm for addition and subtraction. The number bonds in the examples below merely represent the possible thinking process in doing the mental calculation, not a new way of adding or subtracting on paper that she needs to learn. If she is poor at mental math, she can always use the standard algorithm, which works in all situations, rather than spend time trying to decide which paper and pencil method to use depending on the situation. Then you can slowly teach her these strategies and provide practice as you continue with the curriculum, or you can use portions of earlier levels of *Primary Mathematics* to teach these strategies.

Some mental math strategies taught in *Primary Mathematics* 1A-3B:

- Add two 1-digit numbers whose sum is greater than 10 by making a 10. (This strategy is useful for students who know the addition and subtraction facts through 10, but have trouble memorizing some of the addition and subtraction facts through 20.)

7 + 5 = 10 + 2 = 12
  /\
 3 2

7 + 5 = 10 + 2 = 12
/\
2 5

- Add tens or hundreds using the same strategies used for adding ones.

$$70 + 50 = 120$$
$$700 + 500 = 1200$$

- Add tens by just adding the tens or by making the next hundred.

22 + 60 = 80 + 2 = 82 (22 split into 2 and 20)

480 + 50 = 500 + 30 = 530 (50 split into 20 and 30)

- Add a 1-digit number when there is no renaming by simply adding the ones together.

45 + 2 = 40 + 7 = 47 (45 split into 40 and 5)

3645 + 2 = 3640 + 7 = 3647 (3645 split into 3640 and 5)

- Add a 1-digit number where adding the ones results in a number greater than 10

⇒ by making the next 10,

67 + 5 = 70 + 2 = 72 (5 split into 3 and 2)

1467 + 5 = 1470 + 2 = 1472 (5 split into 3 and 2)

⇒ or by using basic addition facts.

67 + 5 = 60 + 12 = 72 (67 split into 60 and 7)

1467 + 5 = 1460 + 12 = 1472 (1467 split into 1460 and 7)

- Add a 2-digit number to a 2-digit number

⇒ by adding the tens and then the ones, using the strategies already learned,

$$48 + 36 = 48 + 30 + 6 = 78 + 6 = 84$$

⇒ or, if one of the numbers is close to a ten, by adding the next ten and then subtracting an appropriate number of ones.

$$48 + 36 = 36 + 48 = 36 + 50 - 2 = 86 - 2 = 84$$

- Subtract ones from tens by recalling number bonds for tens.

10 – 7 = 3

80 – 7 = 73 (80 split into 70 and 10 – 7)

4380 – 7 = 4373 (4380 split into 4370 and 10 – 7)

- Subtract a 1-digit number when there are enough ones by simply subtracting the ones.

67 – 2 = 65 (67 split into 60 and 7)

8867 – 2 = 8865 (8867 split into 8860 and 7)

- Subtract a 1-digit number when there are not enough ones

⇒ by subtracting from a 10,

85 – 7 = 5 + 73 = 78 (85 split into 5 and 80)

385 – 7 = 300 + 5 + 73 = 378 (385 split into 305 and 80)

⇒ or by using basic subtraction facts.

$85 - 7 = 70 + 8 = 78$

70 15

$385 - 7 = 370 + 8 = 378$

370 15

- Subtract tens or hundreds using the same strategies used for subtracting ones.

$150 - 70 = 80$
$850 - 70 = 780$
$8500 - 700 = 7800$

- Subtract a 2-digit number

⇒ by subtracting the tens and then the ones, using the strategies already learned,

$85 - 47 = 85 - 40 - 7 = 45 - 7 = 38$

⇒ or, if one of the numbers is close to a ten, by subtracting the next ten and then adding back in an appropriate number of ones.

$85 - 47 = 85 - 50 + 3 = 35 + 3 = 38$

Students can use these same strategies when adding or subtracting larger numbers where there are only one or two non-zero digits. For example:

⇒ 670,000 + 50,000
67 ten thousands + 5 ten thousands = 72 ten thousands (67 + 5 = 72)
so, 670,000 + 50,000 = 720,000

⇒ 85,000,000 – 7,000,000
85 millions – 7 millions = 78 millions (5 + 73 = 78)
So, 85,000,000 – 7,000,000 = 78,000,000

In earlier levels of *Primary Mathematics*, students learned to use a part-whole model for addition and subtraction as a tool for determining what equation to use to solve word problems or equations with a missing addend, minuend, or subtrahend. A part-whole model for multiplication and division can also be used to remind your student of the relationship between multiplication and division, and to solve equations with a missing factor, divisor, or dividend.

Part-whole model for addition and subtraction:

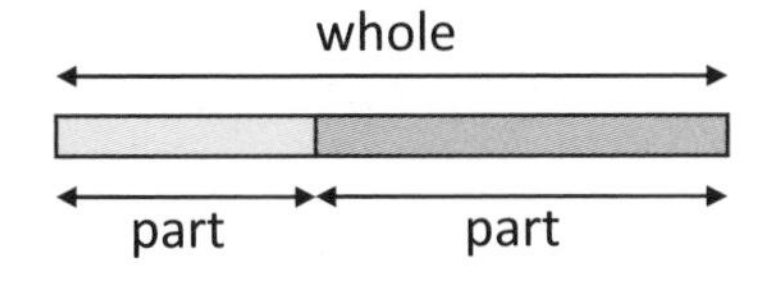

⇒ Given two parts, we can find the whole by addition.

⇒ Given a whole and a part, we can find the other part by subtraction.

Part-whole model for multiplication and division:

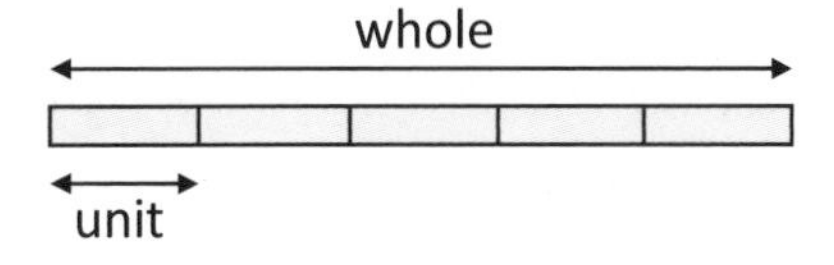

⇒ Given the whole and the number of units, we can find the value of a unit by division.

⇒ Given the value of a unit and the number of units, we can find the whole by multiplication.

If your student did not use earlier levels of *Primary Mathematics*, you can use the models to help him solve problems with missing numbers. If he has used earlier levels of *Primary Mathematics*, he will probably not need to draw the diagrams.

In previous levels the missing number was represented by a blank line. In this chapter, your student will be introduced to representing an unknown number with a letter. For example:

⇒ 32,000 – ___ = 6000 or $32{,}000 - n = 6000$

32,000 is the whole, and 6000 is a part. If we draw a part-whole model for this, we can easily see that we can subtract 6000 from 32,000 to get the missing number.

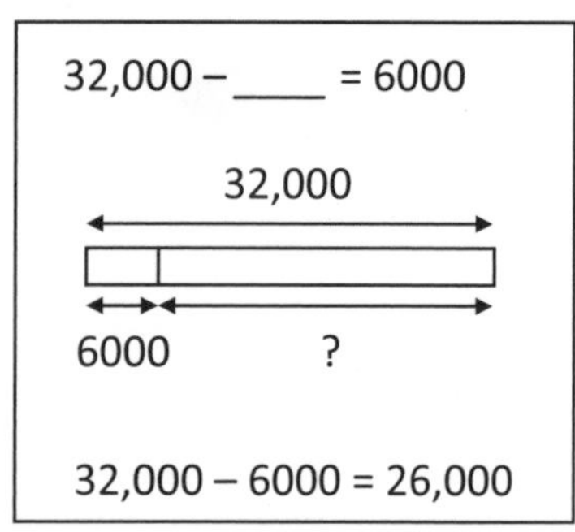

⇒ 4 x ___ = 360,000 or $4 \times a = 360{,}000$

360,000 is the whole. 4 could be considered the number of units. If we draw a model for this, we can see that we can divide to find the missing number.

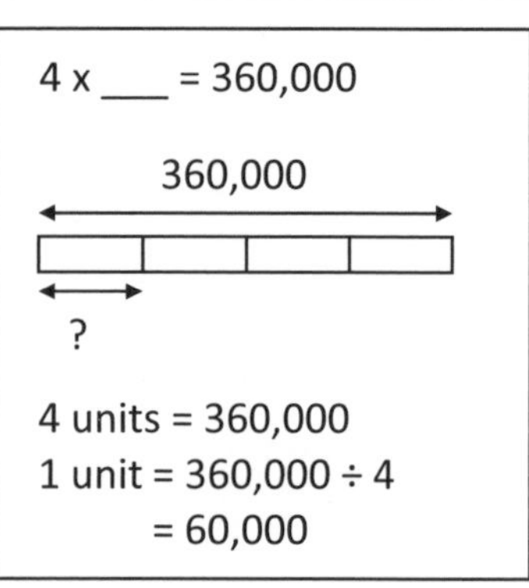

(1) Understand, read, and write numbers within 1 million

Activity

Discuss base-10 concepts through 1,000,000. You can use the following suggested discussion, writing numbers on a place-value chart and adding a new column to the chart every time a new place is needed.

Ask : How many digits are there? (10, including 0)

We can write the numbers up to 9, but to write the number for one more than 9, we have to add a new place to the left, the **tens place**.

Ask: What does the 1 in the tens place mean? It means that we have a group of ten ones.

Now we can count to 99, writing the next digit in the tens place every time we get to ten more ones.

Ask: How many tens and ones there are in 99? (9 tens and 9 ones) After 99, we need a new place, called the **hundreds place**.

Ask: What does the 1 in the hundreds place mean? It means that we have a group of ten tens. It also means we have a hundred ones.

Now we can keep changing digits to write numbers up to 999, writing the next digit in the tens place every time we have ten ones and the next digit in the hundreds place every time we have ten tens.

Ask: What do we do to write one more than 999? We make another place value, called the **thousands place**, and write a 1 there to stand for a group of ten hundreds.

Ask: How many hundreds are there in a thousand? (10) How many tens are there in a thousand? (100)

Now we can write numbers up to 9 thousands, 9 hundreds, 9 tens, and 9 ones, or nine thousand, nine hundred ninety-nine. After 9999, we have to make another place. This is called the **ten-thousands place**.

Ask: How many thousands are in ten thousand? (10) How many hundreds are in ten thousand? (100) How many tens are in ten thousand? (1000)

After 99,999 (ninety-nine thousand, nine hundred ninety-nine) we need another place. This is called the **hundred-thousands place**.

Ask: How many ten thousands are in a hundred thousand? (10) How many thousands are there in a hundred thousand? (100)

We can continue to count for a long time, to 999,999 (nine hundred ninety-nine thousand, nine hundred ninety-nine). To add one more, we need another place. This is called the **millions place**.

Tell your student that after every three places from the right, we put a comma. When we have only 4 digits, the comma is optional.

1
2
...
9
10
11
...
19
20
...
99
100
101
...
110
111
...
200
...
999
1,000
...
9,999
10,000
10,001
...
99,997
99,998
99,999
100,000
100,001
...
100,100
...
101,100
...
110,000
...
999,999
1,000,000

Discussion

Concept pages 8-9

If you have base-10 blocks available, you can show your student the thousand-cube and ask him to imagine that the unit-cube is made up of 1000 units, or is like the thousand-cube. Then the ten-rod would be ten thousand cubes, the hundred-flat one hundred thousand cubes, and the original thousand-cube is now one million cubes.

Task 1, p. 10

Get your student to read the numbers on p. 10 out loud. Point out that we read numbers in sets of three, as hundreds, tens, and ones of thousands, and then as hundreds, tens, and ones of ones. So we read the 23 before the comma the same way as we would read the number 23, but add the word "thousands" since it is 2 ten thousands and 3 one thousands, or twenty-three thousands.

Tasks 2-3, p. 11, and Tasks 5-6, p. 12

2. 2; 3; 5; 4; 6
3. (a) 3 (b) 4

Practice

Task 4, p. 11, and Tasks 7-10, pp. 12-13.

4. (a) 20,800 (b) 35,062
   (c) 88,070 (d) 70,003
   (e) 84,090
5. 1; 2; 4; 9; 3; 6
6. (a) 4 (b) 300; 5
7. (a) 270,600 (b) 572,063
   (c) 300,050 (d) 800,008
   (e) 404,040
8. (a) 8,012 (b) 49,501
   (c) 17,004 (d) 90,090
   (e) 401,062 (f) 970,505
   (g) 700,009
9. (a) three thousand, ninety-six
   (b) seven thousand, two hundred eighty
   (c) five thousand, two
   (d) twenty-seven thousand, one hundred sixty-five
   (e) eighteen thousand, fifty-seven
   (f) forty-two thousand, six hundred five
   (g) thirty thousand, three
   (h) sixty thousand, one hundred nine
   (i) eighty-one thousand, nine hundred
   (j) four hundred thirty-five thousand, six hundred seventy-two
   (k) five hundred thousand, five hundred
   (l) four hundred four thousand, forty
   (m) eight hundred forty thousand, three hundred eighty-two
   (n) six hundred thousand, five
   (o) nine hundred ninety-nine thousand, nine hundred ninety-nine
10. 800,000 + 5000 + 600 + 20

Workbook

Exercise 1, pp. 7-9 (answers p. 21)

(Error 2008 printing: 1(a) discs in thousands place need another 0.)

Enrichment

Have your student write the numbers for these or similar problems. You can either say the problems out loud, or write them down:

⇒ 80 thousands, 500 tens (85,000)

⇒ 60 ones, 600 tens, 600 thousands (606,060)

⇒ 30 ones, 40 tens, 20 hundreds, 30 thousands, and 50 ten thousands (532,430)

⇒ 4100 ones, 63 hundreds, 60 thousands (70,400)

(2) Interpret 6-digit numbers

Discussion

Tasks 11-13, pp. 13-14

If your student has difficulties with place-value concepts, use actual place-value cards and have her illustrate the numbers in these tasks and additional ones using the cards.

11. 8000
    60,000
12. (b) 35,260
    (c) 2
    (d) 3
    (e) 30,000; 5000; 200; 60; 0
13. (b) 345
    (c) 20,000

Practice

Task 14-15, pp. 14-15

Workbook

Exercise 2, pp. 10-12 (answers p. 21)

14. (a) 8
    (b) 600,000
    (c) 0
    (d) 600,000
15. (a) 800 (b) 80,000 (c) 8,000
    (d) 800,000 (e) 8000 (f) 8

Reinforcement

Write some other 5-digit or 6-digit numbers and ask your student what each digit stands for.

Place discs on a place-value chart and ask questions similar to those in Task 12.

Write a 5-digit or 6-digit number and have your student place discs on a place-value chart to represent the number. Ask him for the digit in different places.

Enrichment

Ask your student the following, or similar, questions. Write down the numerals; she does not have to do these orally. Allow her to use place-value cards or discs if needed.

⇒ What is the place value of 0 in 602,432? (ten thousands)

⇒ Find the missing number in 49,703 = 3 + 9,000 + ____. (40,700)

⇒ What is the number 500 + 43,000 + 2? (43,502)

⇒ What is the difference between 64,208 and 4008? (60,200)

⇒ What is the place value of 7 in (read the number out loud or write it in words) four hundred three thousand, one hundred seventy-six? (tens)

⇒ What is the sum of the place values of 4 in 564,464? (1101)

⇒ What is the difference in place values of 5 in 500,456? (99,990)

(3) Understand, read, and write numbers within 1 billion

Activity

Write the number **555,555**.

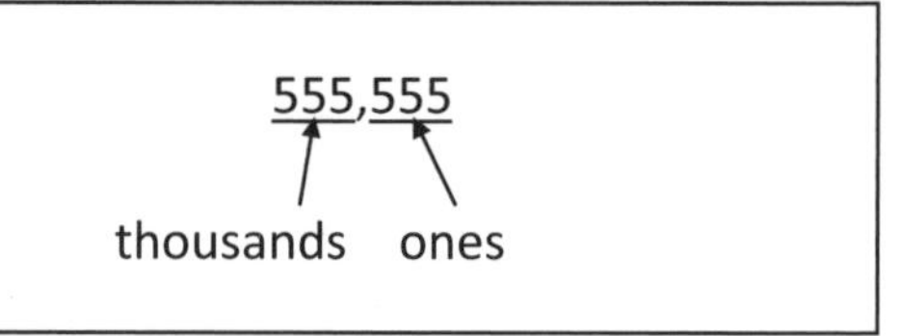

Point to the right-most group of 3. Tell your student that these are places for ones. If we put a 5 in the hundreds place, that tells us there are 5 hundred ones. Point to the next group of 3 digits. Tell him that these are places for thousands. A 5 in the hundreds place in this group means we have 5 hundred thousands.

Tell your student that for larger numbers, we can keep adding place values to the left. The next three places to the left are for millions, ten millions, and hundred millions. Write the number **555,555,555** and ask her to read the number and to identify each place. (Five hundred fifty-five million, five hundred fifty-five thousand, five hundred fifty-five)

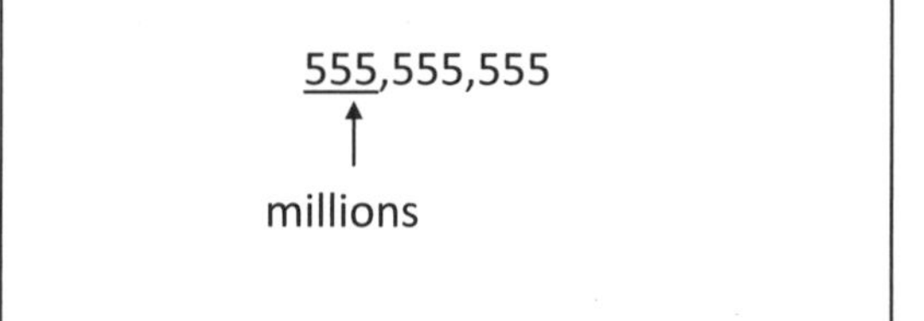

Discussion

Tasks 16-18, pp. 15-16

16. 2000; millions
17. 3
18. 9; 1; 8

Practice

Tasks 19-24, pp. 16-17

19. 400,000,000 + 30,000,000 + 600,000 + 10,000 + 2,000 + 40 + 3
20. 802,109
21. (a) 90,000,000
    (b) millions
22. (a) 9
    (b) 2
    (c) 800,000,000
23. (a) 6,000,000
    (b) 70,003,000
    (c) 42,861,003
    (d) 420,072,130
24. (a) five million
    (b) fourteen million, one hundred twenty-six thousand
    (c) ninety million, forty thousand, three
    (d) four hundred fifty million, one hundred twenty-five thousand, four hundred

Workbook

Exercise 3, pp. 13-14 (answers p. 21)

Enrichment

Your student may be interested in researching the names for the periods in larger numbers. In the U.S. system the names for the periods are based on the number of 000's after 1000. A billion has 2 sets of 000's after a thousand (1,000,000,000), a trillion has 3 sets, a quadrillion 4 sets, and so on. A googol is a 1 followed by 100 zeros, and a googolplex a 1 followed by a googol number of zeros.

Ask your student the problems listed on the next page. Do the first three with him, write the answers out each time so he can see a pattern, and have him read each number out loud. See if he can do the rest on his own. If you want to find the number of hundreds in 1 million, you can simply remove the 0's up to the hundreds place, and adjust the commas.

- ♦ 1,000,000 = _______ ones — 1,000,000 ones; one million ones
- ♦ 1,000,000 = _______ tens — 100,000 tens; one hundred thousand tens
- ♦ 1,000,000 = _______ hundreds — 10,000 hundreds; ten thousand hundreds
- ♦ 1,000,000 = _______ thousands — 1,000 thousands; one thousand thousands
- ♦ 1,000,000 = _______ ten thousands — 100 ten thousands; one hundred ten thousands
- ♦ 1,000,000 = _______hundred thousands — 10 hundred thousands; ten hundred thousands
- ♦ 90,000,000 = _______ hundred thousands — 900 hundred thousands; nine hundred hundred thousands

You may want to do some of the following activities to help give your student an idea of the magnitude of a million. You may have to do the multiplication and division computations for her, since multiplication and division by a 2-digit number or greater than 1 digit has not been taught yet.

How much is a million?

⇒ If pennies were laid side by side, how long would a row of a million pennies be? Put pennies side by side until they are a foot long. 16 pennies in one foot x 5280 feet in 1 mile = 84,480 pennies. Divide 1,000,000 by 84,480. A million pennies side by side is about 12 miles long.

⇒ How much do one million pennies weigh? 1 penny weighs about one tenth of an ounce. So there are 160 pennies in a pound. Divide 1,000,000 by 160. A million pennies weigh about 6250 pounds, about the weight of a half-grown elephant.

⇒ How high would a stack of a million pennies be? There are 16 pennies in a stack an inch high, 16 x 12 = 192 pennies in a foot, 192 x 5280 = 1,013,760 pennies in a mile. So a stack of a million pennies would be about a mile high.

⇒ How long would it take to count to a million, if you counted at the rate of one number per second? There are 60 seconds in a minute, 3600 seconds in an hour, and 86,400 seconds in a day. Divide 1,000,000 by 86,400. It takes about 12 days.

⇒ Page 9 in the textbook shows a million-cube. If each unit-cube is 1 cm on each side, how big is this cube? There are 100 cubes along each side. 1 cube is 1 cm, so 100 cubes = 1 meter. The cube would be one meter on the side.

⇒ If you could live a million days, how old would you be? Divide 1,000,000 by 365 days and the answer is about 2740 years old.

⇒ How many liters are a million drops of water? Use an eye dropper and a medicine spoon and count the number of drops in a milliliter. There are about 20 drops in a milliliter. Multiply by 1000 to get the number of drops in a liter, and divide 1,000,000 by this number to get the number of liters. A million drops is about 50 liters. What could you fill with 50 liters?

(4) Compare numbers and continue a regular number pattern

Discussion

Task 25, p. 17

This task is reproduced on p. 15 of the appendix so that your student can fill in the missing numbers. He needs to determine which digit is increasing or decreasing for each row or column and count on or back by the digit in that place. It is not necessary to read the whole number properly to determine the pattern, just the digit in the place value that is changing and one or two before that. For example, in the second row:
29,500; 29,600; 29,700; ______; ______; ______; 30,100
The hundreds are increasing, so we can read: "two-ninety-five, two-ninety-six, two-ninety-seven, two-ninety-eight, two-ninety-nine, three hundred, three-oh-one." Reading the number in a simplified way allows him to determine the pattern, rather than focusing on reading the number accurately.

25.

5000	6000	7000	**8000**	**9000**	**10,000**	**11,000**
					20,000	
29,500	29,600	29,700	**29,800**	**29,900**	**30,000**	30,100
			28,800		**40,000**	
24,230			**27,800**		**50,000**	
24,130			26,800		60,000	
24,030			**25,800**		70,000	
23,930			24,800			
23,830	23,820	23,810	**23,800**	**23,790**	**23,780**	23,770
23,730						
23,630	**23,640**	23,650	**23,660**	23,670	**23,680**	23,690

Practice

Tasks 26-27, p. 17

Your student can do these by counting on or back in the correct place value. If necessary, use place-value cards or number discs on a place-value chart to help her count on or back by adding or removing discs, renaming as needed.

26. (a) 355,084
    (b) 2,034,006
    (c) 34,567,203
    (d) 199,000,000
    (e) 89,999,600
27. (a) 103,002; 113,002
    (b) 2,742,000; 742,000
    (c) 100,000,000; 102,000,000

Discussion

Task 28, p. 18

Your student needs to find the value of each tick mark. To do this, he can find the difference between the two labeled points and divide that by the number of divisions, or units, between the two labeled points. In 28(a), for example, 500 is the difference between 5000 and 5500, and there are 5 units between 5000 and 5500, so the next mark after 5000 (A) is 5100, and the second mark after 5000 is 5200. In (b), each division from one tick mark to the next is 1000.

28. (a) A: 5100 B: 5300 C: 5700
    D: 5900 E: 6400
    (b) P: 49,000 Q: 52,000 R: 54,000
    S: 58,000 T: 61,000

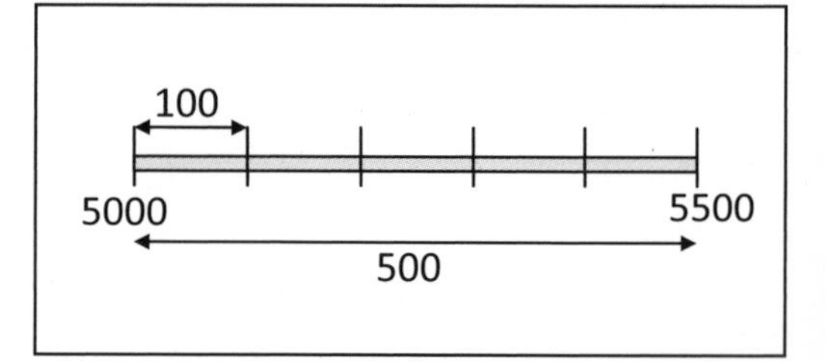

Tasks 29-31, pp. 18-19

The digit in the highest place has the greatest value, so we compare the digits in that place first. In Task 31, point out that the digit in the highest place for the numbers we are comparing isn't always the first digit of both numbers.

29. (a) 56,700
    (b) 32,645
30. 456,978,000
31. 154,030

Remind your student that we use > for **is greater than** and < for **is less than**. Write a few numbers and ask him to put the correct symbol between them.

456,430 ? 456,340
456,340 < 456,430

Practice

Task 32, p. 19

If needed, rewrite the numbers on a place-value chart, aligning the digits.

32. (a) 53,607; 53,670; 53,760; 56,370
    (b) 324,468; 324,648; 342,468
    (c) 2,357,000; 2,537,000; 3,257,000; 4,257,000
    (d) 9,694,400; 61,290,400; 63,290,400; 93,690,900; 96,920,004

Workbook

Exercise 4, pp. 15-16 (answers p. 22)

Reinforcement

Extra Practice, Unit 1, Exercise 1, pp. 7-10

Enrichment

Write a number and give a rule. You can combine rules for different place values for more challenge. Your student writes down the next five numbers according to the rule. For example,

48,300, Rule: + 1000 and – 200

48,300; 49,100; 49,900; 50,700; 51,500; 52,300

Game

Material: 4 sets of number cards, 0-9.

Procedure: Shuffle the deck and place it face down in the middle. Players take turns drawing one card at a time and putting the card in front of him or her to form a 6-digit number. The cards cannot be rearranged once placed. The student with the greatest number wins.

(5) Review: Mental math

Note

This lesson is a review of *some* mental math strategies learned in earlier levels of *Primary Mathematics*. If your student is competent in these strategies, you may wish to skip this lesson.

Activity

Write the expression **67 + 8** and discuss strategies for finding the sum.

⇒ Make the next ten. 67 needs 3 more to get to 70. Split 8 into 3 and 5. Add 3 to 67, then 5 to 70.

⇒ Split the 67 into 65 and 2, and make a ten by adding the 2 to the 8. Add 65 and 10.

⇒ Find the tens first. Look ahead to see that adding the ones (8+7=15) will give another ten. Add 1 to the tens, and then write down 7 for the tens, and 5 for the ones.

⇒ Since 8 is close to 10, add 10 to 67 and then subtract 2.

67 + 8

67 + 8 = 70 + 5 = 75
(8 split into 3 and 5)

67 + 8 = 65 + 10 = 75
(67 split into 65 and 2)

67 + 8 = 60 + 15 = 75

67 + 8 = 67 + 10 – 2 = 75

Provide other problems involving addition of a 1-digit number to a 2-digit number. Include ones that do not require renaming so that your student will have to first determine whether the sum of the ones is more than ten before deciding on a strategy, since with a problem such as 62 + 5 he can simply add ones. You can use the problems in Mental Math 1 in the appendix.

Write the expression **67 + 58** and discuss strategies for finding the sum.

⇒ Add the tens to 67 (6 tens + 5 tens = 11 tens, 60 + 50 = 110, 67 + 50 = 117), and then the ones.

⇒ Add the tens together, and the ones separately, then add the two sums.

⇒ Since 58 is near a ten, add 60 to 67 and then subtract 2.

67 + 58

67 + 58 = 117 + 8 = 125
(58 split into 50 and 8; 8 split into 3 and 5)

67 + 58 = 110 + 15 = 125
(67 split into 60 and 7; 58 split into 50 and 8)

67 + 58 = 67 + 60 – 2
= 127 – 2
= 125

Provide a few other problems involving addition of 2-digit numbers, including ones that don't involve renaming, such as 62 + 55. You can use the problems in Mental Math 2 in the appendix.

Write the expression **73 – 8** and discuss strategies for finding the difference.

73 – 8

⇒ Since there aren't enough ones, subtract 8 from one of the tens. Split 73 into 63 and 10, subtract 8 from the 10, leaving 2, which is added to 63.

73 – 8 = 63 + 2 = 65
/\\
63 10

⇒ Split 73 into 70 and 3, subtract 8 from 70, and add back in the 3.

73 – 8 = 62 + 3 = 65
/\\
3 70

⇒ Split 73 into 60 and 13, and subtract 8 from 13.

73 – 8 = 60 + 5 = 65
/\\
60 13

⇒ Since 8 is near 10, subtract 10 from 73 and then add 2.

73 – 8 = 73 – 10 + 2 = 65

Provide a few other examples, including ones where there is no renaming, such as 78 –2, where your student would simply subtract ones from ones, so that she must first determine whether there are enough ones to subtract from before subtracting from a ten. You can use Mental Math 3 in the appendix.

Write the expression **73 – 38** and discuss strategies for finding the difference.

73 – 38

⇒ Subtract the tens first, and then the ones, using the strategies already learned for subtracting a 1-digit number.

73 – 38 = 43 – 8 = 35
/\\
30 8

⇒ Since 38 is close to 40, subtract 40 and then add 2.

73 – 38 = 73 – 40 + 2 = 35

Provide a few other examples, including ones where there is no renaming, such as 78 – 32. You can use Mental Math 4 in the appendix.

Note

Mental math pages can be done at any time, not just during the lesson in which they are referred to. They are not meant to be done orally; your student can see the written problem and write down the answer. You can have him do each one more than once, such as one mental math exercise each day or every few days for a while, or when the lesson is short.

(6) Use place-value concepts in mental computation

Note

This lesson and the next are primarily review of concepts learned in earlier levels, but with larger numbers, and the use of a letter to stand for an unknown rather than a blank. If your student has not done previous levels of *Primary Mathematics*, you may want to take several days to cover this lesson. If she has done earlier levels of *Primary Mathematics*, you may be able to simply have her do the Tasks 33-35 on textbook p. 19 and see if any review is needed.

Activity

Discuss the addition and subtraction problems shown at the right, using place-value terminology. We can add tens, hundreds, or larger numbers using the same strategies as we used with ones. Underline the zeros as you discuss the computation. We can add 26 thousands + 9 thousands in the same way as 26 ones + 9 ones.

Note that in the last addition example, we add hundreds, since we can only "cover up" two 0's, or hundreds, for one of the numbers. In the last subtraction example, we subtract hundreds, even though 30,000 has four 0's. We can only "cover up" the same number of 0's for each number.

26 + 9 = 35

26,000 + 9000 = ?
26 thousands + 9 thousands = 35 thousands
26,000 + 9000 = 35,000

26,000,000 + 9,000,000 = ?
26 millions + 9 millions = 35 millions
26,000,000 + 9,000,000 = 35,000,000

30,000 + 4100= ?
300 hundreds + 41 hundreds = 341 hundreds
30,000 + 4100 = 34,100

26 – 9 = 17

26,000 – 9000 = ?
26 thousands – 9 thousands = 17 thousands
26,000 – 9000 = 17,000

26,000,000 – 9,000,000 = ?
26 millions – 9 millions = 17 millions
26,000,000 – 9,000,000 = 17,000,000

30,000 – 4100= ?
300 hundreds – 41 hundreds = 259 hundreds
30,000 – 41000 = 25,900

Discuss the multiplication problems shown at the right, using number discs to illustrate if necessary. Just as we would need to triple the 6 one-discs to multiply 6 by 3, so we would triple 6 ten-thousand-discs to multiply 60,000 by 3.

Note that in the last example there are more 0's in the product than in one of the factors; one of the 0's came from the product of 5 and 8.

6 x 3 = 18

60,000 x 3 = ?
6 ten thousands x 3 = 18 ten thousands
60,000 x 3 = 180,000

5 x 800,000 = ?
5 x 8 hundred thousands
= 40 hundred thousands
5 x 800,000 = 4,000,000

Discuss the division problems at the right. Show your student that he can cover up the 0's in the first number until he gets to a number he recognizes as one that can be divided by the second number, if there is one.

Note that in this third example, we can only "cover up" five 0's, not all six 0's.

56 ÷ 8 = ?
56 ones ÷ 8 = 7 ones
56 ÷ 8 = 7

560,000 ÷ 8 = ?
56 ten thousands ÷ 8 = 7 ten thousands
560,000 ÷ 8 = 70,000

3,000,000 ÷ 6 = ?
30 hundred thousands ÷ 6 = 5 hundred thousands
3,000,000 ÷ 6 = 500,000

Note: All these dividends in these problems are easy multiples of the divisor. Students will be rounding to one or two non-zero digits that are multiples of the divisor when they use approximation to estimate the answer to a division problem. You can give your student the example at the right, 34,000 ÷ 5, as an example where taking off 0's still might result in a problem where we need to use the standard division algorithm.

34,000 ÷ 5 = ?
34 is not divisible by 5.
340 ÷ 5 = 68
34,000 ÷ 5 = 6800

Practice

Task 33, textbook p. 19

33. (a) 14,000 (b) 31,000
(c) 9000 (d) 28,000
(e) 50,000 (f) 5000
(g) 27,000,000 (h) 320,000,000

Reinforcement

Mental Math 5-8

(7) Find the missing number

Activity

Write and discuss the following problems.

⇒ 50,000 + ____ = 230,000

Get your student to come up with the related problem, 5 + ___ = 23 (from 5 ten thousands + _____ = 23 ten thousands). Remind her that addition can be thought of as putting two parts together to make a whole. Ask what is missing in the equation 5 + ____ = 23, a part or a whole. Since 23 is the whole, and 5 a part, the equation is missing a part.

Show your student a part-whole model for this problem. From it, we see that we can use subtraction to find the missing part. 23 – 5 = 18. Since the original problem was in ten thousands, the missing number is 180,000.

Tell your student that we can use other symbols than a blank line to represent an unknown number. The textbook uses color patches. A letter can also be used, such as n, instead of a blank line. Then we solve the problem for the *value* of n.

50,000 + ____ = 230,000
5 + ____ = 23

23

5 ?

23 – 5 = 18
50,000 + **180,000** = 230,000

50,000 + n = 230,000
n = 180,000

⇒ 32,000 – n = 6000

Get your student to write the related problem, 32 – n = 6. If he has trouble with using n, erase it, write a blank line, and then after the problem is solved re-do the same problem using n. Ask him what number is the whole and which numbers are the parts. 32 is the whole, and n and 6 are the parts. Show him the model for this problem. From the model, we see that we can solve this by subtracting 6 from 32 to get n. Since the original problem was in thousands, n is 6000.

32,000 – n = 6000
32 – n = 6

32

6 ?

32 – 6 = 26
32,000 – 6000 = 26,000
n = 6000

⇒ s – 120,000 = 41,000

Get your student to tell you which value is the whole. The unknown number is the whole. 120,000 and 41,000 are the parts. You can show this with a model. Have her find the value of s. She needs to add the parts.

s – 120,000 = 41,000

s

120,000 41,000

120,000 + 41,000 = s
120 + 41 = 161
s = 161,000

⇒ $a \times 4 = 360{,}000$

Tell your student that we are multiplying a by 4. Draw the model. Lead him to see that we can make a the value of one unit. We have the total, 360,000, and the number of parts, 4, and want to find the value of each part, a. We divide: 360,000 ÷ 4. Since 36 ÷ 4 = 9, then 360,000 ÷ 4 = 90,000.

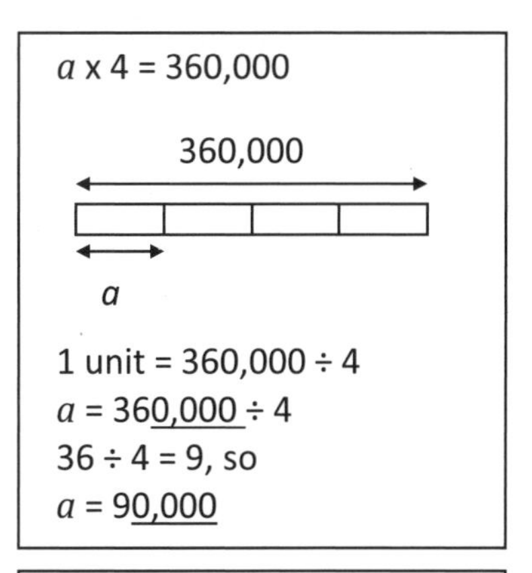

⇒ $n \div 3 = 600$

Ask your student which number is the total. Since this is division, n is the total. Lead her to see that we can interpret the problem as dividing n into 3 parts, each 600. Draw the model. To find n, we multiply.

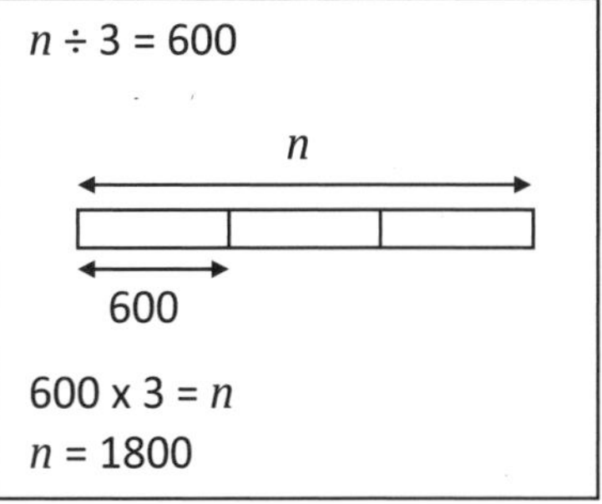

⇒ $4200 \div y = 600$

Get your student to write the related problem, $42 \div y = 6$. Ask him what equation we can use to find the answer. Lead him to see that we can think of this problem as dividing 42 into parts to give 6 in each part. We can find the number of parts by finding the number that when multiplied by 6 gives 42. Since 6 x 7 = 42, then we know that 42 ÷ 7 = 6, and 4200 ÷ 7 = 600.

$4200 \div y = 600$
$42 \div y = 6$
$6 \times y = 42$
$6 \times 7 = 42$
$600 \times 7 = 4200$
$y = 7$

Practice

Tasks 34-35, p. 19

Workbook

Exercise 5, pp. 17-18 (answers p. 22)

34. (a) 34 – 4 = 30 →
    34,000 – 4000 = **30,000**
    (b) 11 – 9 = 2 →
    110,000,000 – 90,000,000 = **20,000,000**
    (c) 36 ÷ ? = 4 → 36 = 4 x ?
    36 = 4 x 9 → 360,000 ÷ **9** = 40,000
    (d) 10 ÷ 5 = 2 →
    100,000 ÷ 5 = **20,000**

35. (a) 63 – 7 = 56 → 63,000 – 7,000 = 56,000
    n = **56,000**
    (b) 44 – 42 = 2 → 4400 – 4200 = 200 →
    4200 + 200 = 4400
    n = **200**
    (c) 72 ÷ 8 = 9 → 720,000 ÷ 8 = 90,000
    n = **8**
    (d) 4 x 7 = 28 → 4 x 7,000,000 = 28,000,000
    n = **7,000,000**

(8) Practice

Practice

Practice A, pp. 20-21

5(c-d): In this curriculum, students have not yet learned to multiply a multi-digit number by more than 1 digit, other than 10. However, your student should be able to answer 5(c) and 5(d) from her experience with place value. If she has trouble with 5 (c), have her think of the 100 as a hundred-disc. Ten of them are ten hundreds, which is 1000. Similarly, for 5(d), 80 thousand-discs is 80 thousands, or 80,000.

Reinforcement

Mental Math 9

Tests

Tests, Unit 1, 1A and 1B, pp. 1-6

1. (a) 11,012
   (b) 115,600
   (c) 700,013
   (d) 880,005
   (e) 5,000,000
   (f) 4,200,000
   (g) 10,000,000
   (h) 8,008,000
   (i) 63,402,600
   (j) 120,004,020

2. (a) two hundred seven thousand, three hundred six
   (b) five hundred sixty thousand, three
   (c) seven hundred thousand
   (d) three million, four hundred fifty thousand
   (e) six million, twenty thousand
   (f) four million, three thousand
   (g) forty-eight million, twenty
   (h) one hundred twenty-six million, three hundred thousand
   (i) nine hundred ninety-nine million, nine hundred ninety-nine thousand, nine hundred ninety-nine

3. (a) 800 (b) 80,000 (c) 80
   (d) 800,000 (e) 8000 (f) 8,000,000

4. (a) 6000 (b) 200,000
   (c) 184,900 (d) 7,609,000
   (e) 9,021,000 (f) 30,000
   (g) 297,000

5. (a) <
   (b) =
   (c) >
   (d) =
   (e) <

6. (a) $a = 100{,}000$
   (b) $a = 4000$
   (c) $a = 8$
   (d) $a = 200$
   (e) $a = 8$

Workbook

Exercise 1, pp. 7-9

1. (a) 4053 (b) 23,405

2. (a) 32,400
   (b) thirty-two thousand, four hundred

3. (a) 8,402 (b) 12,793
   (c) 90,511 (d) 88,008
   (e) 99,999

4. (a) two thousand, eighty
   (b) nine thousand, two hundred fifteen
   (c) forty-seven thousand, ten
   (d) eighty-nine thousand, one hundred two
   (e) forty thousand, nine hundred
   (f) seventy-eight thousand, nine hundred ninety-nine

5. (a) 24,608 (b) 16,011
   (c) 99,009 (d) 312,460
   (e) 802,003 (f) 540,014
   (g) 900,909

6. (a) fifty thousand, two hundred thirty-four
   (b) twenty-six thousand, eight
   (c) seventy-three thousand, five hundred six
   (d) three hundred sixty-seven thousand, four hundred fifty
   (e) five hundred six thousand, nine
   (f) four hundred thirty thousand, sixteen
   (g) eight hundred thousand, five hundred fifty

7. 900,000 + 50,000 + 7000 + 5

Exercise 2, pp. 10-12

1. (a) 9, 20, 500, 3000, 20,000
   (b) 8, 10, 600, 0, 40,000
   (c) 3, 20, 0, 5000, 40,000
   (d) 8, 80, 800, 8000, 80,000

2. (a) 70,000
   (b) 2; 200
   (c) 4; 8

3. (a) 4000
   (b) 4
   (c) 500

4. (a) 3
   (b) 6000
   (c) 40,000
   (d) 2000
   (e) 100

5. (a) 4307
   (b) 56,400
   (c) 30,768
   (d) 11,400
   (e) 90,090

6. (a) 7000 (b) 6
   (c) hundreds (d) 40

7. (a) 42,108 (b) 562,032
   (c) 770,077 (d) 900,214

8. (a) 800 (b) 300,000
   (c) 3000 (d) 8

Exercise 3, pp. 13-14

1. (a) 3,000,000 (b) 4,150,000
   (c) 6,031,000 (d) 7,208,000
   (e) 5,005,000 (f) 95,909,000
   (g) 710,000,000

2. (a) four million
   (b) six million, three hundred fifty thousand
   (c) three hundred eight million, five hundred sixty-seven thousand
   (d) seventeen million, seven hundred three thousand
   (e) three million, forty thousand
   (f) five million, six thousand
   (g) one hundred forty-nine million, ninety-nine thousand

3. $2,003,705
   two million, three thousand, seven hundred five dollars

4. $2,400,000
   two million, four hundred thousand dollars

Workbook

Exercise 4, pp. 15-16

1. (a) 43,628
   (b) 253,240
   (c) 89,900
   (d) 86,100,000
   (e) 100
   (f) 10,000
   (g) 1000
   (h) 1000

2. (a) 526
   (b) 30,000

3. (a) 36,552; 37,552
   (b) 71,880; 72,080
   (c) 303,610; 313,610

4. (a) 31,862
   (b) 42,650
   (c) 33,856
   (d) 65,703

5. (a) 3695; 3956; 30,965; 35,096
   (b) 296,870; 435,760; 462,540; 503,140

Exercise 5, pp. 17-18

1. (a) 16,000
   (b) 37,000
   (c) 24,000

2. (a) 9,000
   (b) 34,000
   (c) 24,000

3. (a) 6000
   (b) 48,000
   (c) 63,000

4. (a) 2000
   (b) 12,000
   (c) 3000

5. (a) 41 – 29 = 12, so
   41,000 – 29,000 = 12,000
   n = **12,000**
   (b) 70 – 24 = 46, so
   70,000 – 24,000 = 46,000
   n = **46,000**
   (c) 54 – 33 = 21, so
   54,000 – 33,000 = 21,000
   n = **21,000**
   (d) whole - part = part, so
   part + part = whole
   24 + 16 = 40, so
   24,000 + 16,000 = 40,000
   n = **40,000**
   (e) 4 x ? = 12
   4 x 3 = 12, so
   40,000 x 3 = 120,000
   n = **3**
   (f) 40 ÷ 5 = 8, so
   40,000 ÷ 5 = 8000
   n = **8000**
   (g) 15 ÷ 5 = 3, so
   15,000 ÷ 5 = 3000
   n = **5**
   (h) 7 x 8 = 56, so
   7000 x 8 = 56,000
   n = **56,000**

Chapter 2– Approximation

Objectives

- Round numbers within 1 billion to the nearest ten, hundred, thousand, ten thousand, hundred thousand, or million.

Vocabulary

- Approximately
- Rounded

Material

- Number lines (appendix p. a16)

Notes

In *Primary Mathematics* 3A, students learned to round numbers of up to 4 digits to the nearest ten, hundred, or thousand. In this chapter, your student will round larger numbers to the nearest hundred, thousand, ten thousand, hundred thousand, and million. He should then be able to round any number to any given place value.

Rounding numbers will be used to estimate the answers to addition, subtraction, multiplication, or division problems. There are times when estimation is the goal, like wanting to know the approximate amount of money needed. Here and in later grades, students will use rounding to estimate an answer in order to determine if their calculated answer makes sense, particularly with regard to place value. This will be particularly useful to check answers when multiplying or dividing by more than 1 digit and in operations with decimals.

By convention, if a number is exactly halfway between the place the number is being rounded to, it is rounded to the higher number. For example, 465 rounded to the nearest ten is 470.

To round a number to a specified place, we look at the digit in the next lower place. If it is 5 or greater than 5, we round up. If it is smaller than 5, we round down. In the examples shown below, we are rounding each of the following numbers to the nearest ten thousand.

248,453 → 250,000   8 in the thousands place; the number is closer to 250,000 than 240,000.

1,874,555 → 1,870,000   4 in the thousands place; the number is closer to 1,870,000 than 1,880,000.

95,000 → 100,000   5 in the thousands place; round up.

When asked to round a number to a specific place value, your student should round the whole number to that place value, rather than rounding to each place value successively. For example, 2446 when rounded to the nearest hundred is 2400, but when rounded first to the nearest ten (2450) and then to the nearest hundred is 2500.

(1) Round numbers

Discussion

Concept page 22

For each number line, ask your student to determine the size of the division between tick marks, the value each tick mark represents, and explain why the numbers are located where the arrow is pointing. In the first number line, the divisions are 10 and in the second they are 100.

Tell your student that when we round a number to a particular place value, we look at the number in the next lower place. When we round 4865 to the nearest hundred, we look at the digit in the tens place. The 6 in the tens place tells us that the number is closer to 4900 than 4800. When we round to the nearest thousand, we look at the digit in the hundreds place. The 8 in the hundreds place tells us that the number is closer to 5000 than to 4000.

> Round 4865 to the nearest hundred.
> 4 8 (6) 5
> ↑
> hundreds place
> 6 is in the tens place; we round up to 4900.

Ask your student to locate 4836 on both number lines and round to the nearest hundred and then the nearest thousand.

Point to 4850 on the first number line. Tell your student we want to round it to the nearest hundred. This number is the same distance from 4800 as 4900. Tell her we will round up when the digit in the next lower place is a 5.

Discussion

Tasks 1-6, p. 23

For Task 4(b), point out that if we round 6047 to the ten and then to the hundred, we get a different answer (6100) than if we simply round to the nearest hundred (6000). Tell your student that he should always round in a single step.

> 1. 490
> 2. (a) 600 (b) 800 (c) 1000
> 3. 5700
> 4. (a) 3700 (b) 6000 (c) 5000
> 5. 17,000
> 6. (a) 23,000 (b) 55,000 (c) 40,000

Discussion

Top of p. 24
Tasks 7-9, p. 24

> 7. (a) 50,000 (b) 70,000
> (c) 800,000 (d) 130,000
> 8. 72,800,000
> 9. 20,000,000

Practice

Task 10, p. 24

> 10. (a) 3,000,000 (b) 1,000,000 (c) 10,000,000
> (d) 44,000,000 (e) 39,000,000 (f) 48,000,000
> (g) 189,000,000 (h) 100,000,000 (i) 100,000,000

Workbook

Exercise 6, pp. 19-20 (answers p. 36)

Reinforcement

Locating numbers on a number line between tick marks requires approximation, because the student must decide which of the marked values the number is closest to. Use the set of number lines on p. a16 of the appendix and ask your student to locate some numbers on them.

(2) Practice

Practice

Practice B, p. 25

1. (a) 70 (b) 660 (c) 1290
2. (a) 300 (b) 1300 (c) 20,800
3. (a) 7000 (b) 11,000 (c) 125,000
4. (a) 8,800,000 (b) 7,600,000 (c) 91,500,000
5. $800
6. $70,000
7. 1,000,000 km
8. (a) 7,000,000 (b) 21,000,000
   (c) 156,000,000 (d) 350,000,000
9. hundred thousands
10. Smallest: 5,348,500
    Greatest: 5,349,499

Reinforcement

Have your student solve the following problems.

⇒ What number is the largest 6-digit whole number with 6 in the thousands place and 4 in the tens place? (996,949)

⇒ What number is the largest whole number that can become 60,000 when rounded to the nearest ten thousand? (60,499)

⇒ What number is the smallest whole number that can become five million when rounded to the nearest hundred thousand? (4,950,000)

⇒ What number is the smallest whole number that can become five million when rounded to the nearest hundred? (4,999,950)

Extra Practice, Unit 1, Exercise 2, pp. 11-12

Test

Tests, Unit 1, 2A and 2B, pp. 7-10

Note: In the first printing of the tests book (2008) the tests for this chapter include problems on estimation. Estimation was covered in *Primary Mathematics* 3, but has not been reviewed and applied to larger numbers at this level yet. It will be reviewed in the next unit. Therefore, you may want to omit these tests, save them for later, or omit problems 5-7 in Test A and problems 7 -10 in Test B and use them after Chapter 2 of Unit 2.

Chapter 3 – Factors

Objectives

- Understand factors, and prime and composite numbers.
- Find the factors of numbers within 100.
- Find the common factors of two numbers within 100.
- Understand that whole numbers can break down into factors in different ways.
- Find a missing factor in equations with two or more factors.

Vocabulary

- Product
- Factor
- Composite number
- Prime factor
- Common factor

Material

- Multilink cubes
- Graph paper (appendix p. a17)
- Number cubes, 2 (labeled 0-5 and 4-9)
- Factor game board (appendix p. a18)

Notes

Factors and multiples are important concepts for working with fractions. In this chapter, your student will learn the term **factor**, how to find the factors of any whole number within 100, and how to find factors common to two numbers. She will also learn the definition of a **prime number** and a **composite number**. Your student needs to be able to easily recall multiplication and division facts before starting this chapter. If necessary, spend some time reviewing the multiplication and division facts through 10 x 10.

A **factor** is a whole number. We will only consider positive factors at this level. Any whole number can be expressed as the product of two or more factors. For example, since 4 x 3 = 12, then 4 and 3 are factors of 12. If a number has two or more factors, the smallest is 1 and the largest is the number itself. Numbers which have only two different factors, 1 and the number itself, are called **prime numbers**. Numbers with more than two factors are called **composite numbers**. The number 1 is neither a prime number nor a composite number.

We can use division to test if one number is a factor of another number, because there is no remainder when a number is divided by one of its factors. For example, 4 is a factor of 12 because 12 ÷ 4 = 3 with no remainder.

Your student will learn to systematically find the factors of a number by checking each number from 1 up until the factors repeat themselves. For example, to find the factors of 24, he can try dividing 24 by 2, 3, 4, 5, and 6. 2, 3, 4, and 6 are factors of 24; 5 is not. The other factor with 6 is 4, so we know all the factors have been found. The factors of 24 are 1, 2, 3, 4, 6, 8, 12, and 24.

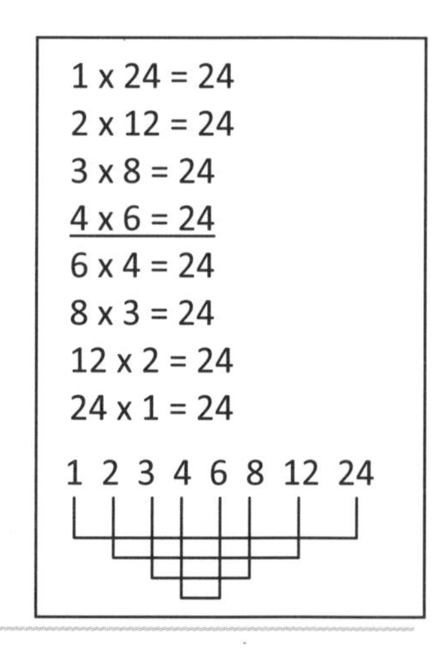

A **common factor** of two numbers is any factor that both numbers have. For example, 4 is a common factor of 12 and 24.

Finding the greatest common factor will be covered in *Primary Mathematics* 5.

In the next chapter, your student will learn about multiples and look for patterns in the multiples of 2, 3, 4, 6, and 9. After this, she will be able to apply rules of divisibility to determine whether a number might be a factor of another number.

In *Primary Mathematics* 3, students learned to multiply more than two numbers together, and that multiplication can be done in any order. For example, 2 x 4 x 5 = 5 x 4 x 2. In this chapter, your student will learn to express a number as the product of two or more factors. For example, 24 = 4 x 6 = 4 x 2 x 3. In *Primary Mathematics* 5, he will learn to express a number as a product of its **prime factors**.

(1) Understand factors by using rectangular arrays

Activity

Give your student 24 multilink cubes. Have her make as many different arrays with them as she can and record the results on graph paper. There are four possible arrays (the orientation does not matter). Write two multiplication equations for each.

1 x 24 = 24
24 x 1 = 24

2 x 12 = 24
12 x 2 = 24

3 x 8 = 24
8 x 3 = 24

4 x 6 = 24
6 x 4 = 24

The factors of 24: 1, 2, 3, 4, 6, 8, 12, 24

Remind your student that the answer to a multiplication problem is called the **product**. Tell him that the numbers that are multiplied by each other are called **factors**. A factor of 24 is a whole number that can be multiplied by another whole number to get the product 24. List the factors of 24: 1, 2, 3, 4, 6, 8, 12, 24. Ask him if 5 is a factor of 24. It is not. We cannot make an array of 24 with 5 on the side.

Give your student 5 cubes and ask her to make as many arrays as she can and list the factors of 5. She can only make one array, so 5 has only two factors, 1 and 5.

Discussion

Concept page 26

Point out that each of the factors in the second equation, 2 x 3 x 4 = 24, can be multiplied by a single other number to give 24. 2 x 12 = 24, 3 x 8 = 24, and 4 x 6 = 24.

Tasks 1-5, pp. 27-28

For Task 2, you can give your student 16 cubes so that he can see if there are any other factors of 16 besides 1, 2, 4, 8, and 16.

Make sure your student understands the definition of a prime number and a composite number. Point out that 1 is neither a prime number nor a composite number, since it has only one factor.

For Tasks 4-5, allow your student to use multilink cubes to find the answers.

1. 3, 6
2. 4 (1 and 16 are also factors of 16.)
4. (a) 1, 7 prime
   (b) 1, 3, 9
   (c) 1, 3 prime
   (d) 1, 2, 3, 6, 9, 18
   (e) 1, 11 prime
   (f) 1, 3, 5, 15
   (g) 1, 2, 5, 10
   (h) 1, 13 prime
5. (a) 8, 10, 24
   (b) 15, 20, 25

Workbook

Exercise 7, pp. 21-22 (answers p. 36)

In problem 5, all your student needs to do to determine if a number is not prime is to think of a number other than 1 and the number itself that is a factor. So 16, 27, and 21 are easily eliminated. She can then test various numbers as factors to determine that 13 and 19 are prime numbers. Or she can use multilink cubes.

(2) Find factors and common factors

Discussion

Tasks 6-8, p. 29

There is no remainder if a number is divided by its factor. A number that is a factor of another number is a common factor of both numbers.

6. (a) yes
   (b) no
7. (a) yes
   (b) no
8. (a) yes
   (b) yes
   (c) yes

Practice

Tasks 9-10, p. 29

9. (a) yes
   (b) yes
10. (a) 4 (b) 8
    (c) 9 (d) 9
    (e) 7 (f) 8

Activity

Write the number **60** on the board and show your student how to systematically find all the factors. Test each number, starting with 2, until the factors start repeating themselves.

- ⇒ 1 and 60 are factors of 60.
- ⇒ 60 ÷ 2 = 30, so 2 and 30 are factors.
- ⇒ 60 ÷ 3 = 20, so 3 and 20 are factors.
- ⇒ 60 ÷ 4 = 15, so 4 and 15 are factors.
- ⇒ 60 ÷ 5 = 12, so 5 and 12 are factors.
- ⇒ 60 ÷ 6 = 10, so 6 and 10 are factors.
- ⇒ 60 ÷ 7 = 8 R 4, so 7 is not a factor.
- ⇒ 60 ÷ 8 = 7 R 4, so 8 is not a factor.
- ⇒ 60 ÷ 9 = 6 R 6, so 9 is not a factor.
- ⇒ The next number to test is 10, but we have already found 10 as a factor, since 6 x 10 = 60. So we have already found all the factors.
- ⇒ The factors of 60 are 1, 2, 3, 4, 5, 6, 10, 12, 15, 20, 30, and 60.

1 x 60 = 60
2 x 30 = 60
3 x 20 = 60
4 x 15 = 60
5 x 12 = 60
6 x 10 = 60

10 x 6 = 60

1, 2, 3, 4, 5, 6, 10, 12, 15, 20, 30, and 60 are factors of 60.

Ask your student to find and list the factors of 48. List the factors for 60, and ask him to circle common factors.

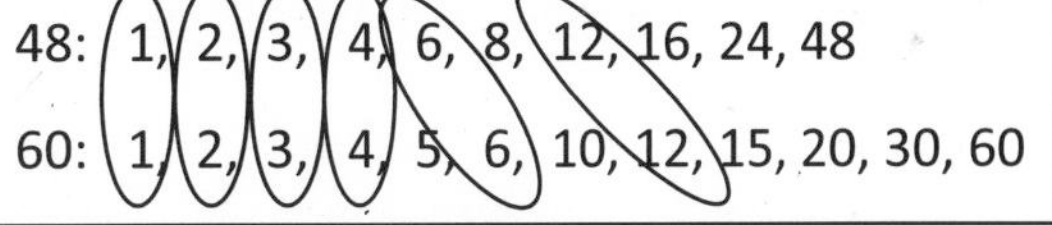

1, 2, 3, 4, 6, and 12 are the common factors of 48 and 60. Ask her which is the largest common factor. (12)

Practice

Tasks 11-14, p. 30

11. 4, 8, 16, 32
12. 24; 16; 12; 8
    8, 12, 16, 24
13. 1, 2, 4, 5, 10, 20, 25, 50, 100
14. (a) 1, 2, 4, 5, 8, 10, 20, 40
    (b) 1, 2, 5, 10, 25, 50
    (c) 1, 3, 5, 15, 25, 75
    (d) 1, 2, 4, 5, 8, 10, 16, 20, 40, 80

Workbook

Exercise 8, pp. 23-24, #1-4 (answers p. 36)

(3) Factor numbers in different ways

Activity

Ask your student to find the common factors of 12 and 48.

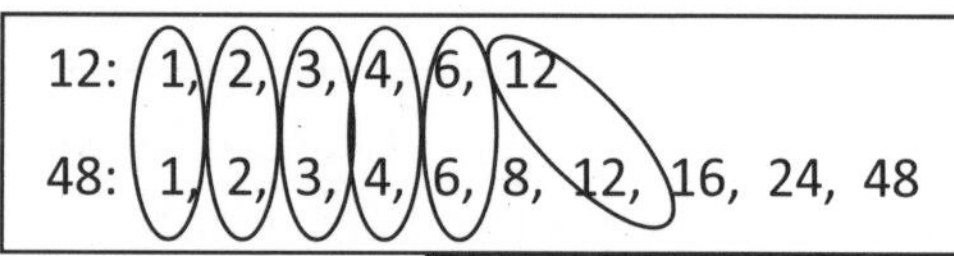

Point out that all the factors of 12 are also factors of 48. Since 12 is a factor of 48, the factors of 12 are also factors of 48.

Write the equation **48 = 12 x 4**. Below it, write an equation replacing 12 with 6 x 2. Ask your student if it is a true statement. It is, since 6 x 2 = 12, and 12 x 4 = 48. Now write **48 = 4 x 2 x 6**. Ask him if this is also true. It is, since multiplication can be done in any order.

48 = 12 x 4
48 = 6 x 2 x 4
48 = 4 x 2 x 6
48 = 8 x 6

Write **12 x 4 = 8 x ___**. Ask your student for suggestions on how we could find the missing number without having to find the product of 12 and 4. We can think about factors of 12 and 4 to get a combination that equals 8. The product of any remaining factors would then be the missing number.

12 x 4 = 8 x ___
2 x 6 x 4 = 8 x ___
2 x 4 x 6 = 8 x 6
8

Practice

Tasks 15-16, p. 31

15. 12 = 2 x **3** x 2
    24 = **4** x 3 x 2
    24 = 2 x **2** x 3 x 2
    24 = 12 x **2**
    24 = 6 x **2** x 2
    24 = 3 x **2** x 2 x 2
    24 = 3 x 2 x **4**
    24 = 3 x **8**
16. (a) 3 (b) 2
    (c) 7 (d) 6
    (e) 12 (f) 32

Workbook

Exercise 8, pp. 25, #5-6 (answers p. 36)

Reinforcement

Extra Practice, Unit 1, Exercise 3, pp. 13-14

Write a number, such as 72, and guide your student in factoring it completely, two factors at a time (other than 1 and the number itself), using a branching structure. An example is shown at the right. Ask her if she notices anything about the numbers at the ends of the branches. They are prime numbers. Ask her to use these numbers to write 72 as the product of more than two factors. Point out that she can do this by combining the prime factors in different ways.

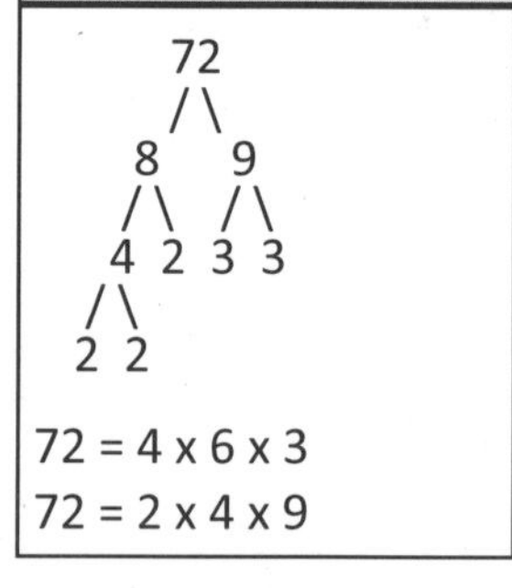

Tests

Tests, Unit 1, 3A and 3B, pp. 11-14

Factor game

Material: A game board of 4 x 5 squares with the numbers 2 through 15 and two stars (appendix p. a18). Two number cubes, one with 0-5 and the other 4-9. A different kind of marker for each player.

Procedure: Players take turns throwing the number cubes to form a two-digit number. The player then puts his marker on an unoccupied number on the game board that is a factor of the number he formed. If the number is a prime number, he puts his marker on a star. The first player with 3 in a row wins.

Chapter 4 – Multiples

Objectives

- Understand the meaning of multiples and relate multiples to factors.
- Determine if a number is a multiple of another.
- List multiples of numbers within 10.
- Learn divisibility rules for 2, 3, and 5.
- Find common multiples of two numbers.

Vocabulary

- Multiple
- Common multiple
- Divisible

Material

- Hundred-chart (appendix p. a19)
- Number cube 1-6

Notes

A **multiple** is any number which is the product of the given number and a whole number. For example, 10 is a multiple of the given number 2 since 2 x 5 = 10. That is, if 10 is a multiple of 2, then 2 is a factor of 10. So when we determine that 2 is a factor of 10, then we can say at the same time that 10 is a multiple of 2.

We can find multiples of a number by simply counting by that number, such as counting by 3's to find multiples of 3.

Your student will be finding the common multiples of two numbers in this chapter by listing the multiples of both, and comparing them. The first **common multiple** is called the least common multiple. Finding the least common multiple is not emphasized at this point in this curriculum, though you may point out the least common multiple when appropriate. A simple way to find a common multiple of both numbers is to simply multiply them together. Your student will need to be able to find common multiples to work with fractions.

In this chapter your student will look at patterns for multiples of the prime numbers 2, 3, and 5. This guide will also have her look for patterns for multiples of 6 and 9. From these patterns we can devise divisibility rules. These rules can help her decide whether a number is **divisible** by another number, that is, whether it divides evenly with no remainder.

A number is divisible by	If
2	The ones digit is 0, 2, 4, 6, or 8.
3	The sum of the digits of the number is divisible by 3.
5	The ones digit is 0 or 5.
6	The number is divisible by both 2 and 3.
9	The sum of the digits of the number is divisible by 9.
10	The ones digit is 0.

Additional notes

The divisibility rules also apply to numbers greater than 100. For your information, some explanations are given below, as well as some divisibility rules for 7 and 8. This is for your information; you do not have to teach this to your student.

⇒ **A number is divisible by 2 if the last digit is 0, 2, 4, 6, or 8.**

2, 4, 6, and 8 have 2 as a factor. Multiples of 10 are divisible by 2, since 10 is. 356 = (3 x 100) + (5 x 10) + 6. Each of the parts is divisible by 2, so the whole number is. 357, on the other hand, is not divisible by 2. 357 = (3 x 100) + (5 x 10) + 7, and the 7 is not divisible by 2. So we only have to look at the last digit to determine if a number is divisible by 2.

⇒ **A number is divisible by 5 if the last digit is 5 or 0.**

10 and every multiple of ten are divisible by 5. 455 = (4 x 100) + (5 x 10) + 5. Each of the parts is divisible by 5, so the whole number is. However, 456 is not divisible by 5, since in (4 x 100) + (5 x 10) + 6, the last part, the 6, is not divisible by 5. We only need to look at the last digit to determine if a number is divisible by 5.

⇒ **A number is divisible by 3 if the sum of the digits is divisible by 3.**

To understand the divisibility rule for 3, we can also divide the number into parts, but we don't use tens. We can write 400 as (4 x 99) + 4 and 50 as (5 x 9) + 5. 456 = (4 x 99) + 4 + (5 x 9) + 5 + 6 = (4 x 99) + (5 x 9) + 4 + 5 + 6. The first two parts, 4 x 99 and 5 x 9, are divisible by 3, because 99 and 9 are. 4 + 5 + 6 = 15, which is divisible by 3.

⇒ **A number is divisible by 9 if the sum of the digits is divisible by 9.**

By the same reasoning applied above, 456 is not divisible by 9, since 4 + 5 + 6 is not divisible by 9. However, 459 is divisible by 9. We can divide it up the same way: 459 = (4 x 99) + 4 + (5 x 9) + 5 + 9 = (4 x 99) + (5 x 9) + 4 + 5 + 9 = (4 x 99) + (5 x 9) + 18. All three parts, including the 18, are divisible by 9, so all we need to be concerned about is the last part, which is the sum of the digits. The same type of argument works for all numbers, so a number is divisible by 9 if the sum of the digits is.

⇒ **A number is divisible by 4 if the last two digits are divisible by 4.**

10 is not divisible by 4, but 100, 1000, etc. are. So we can expand a number to the tens place and just check the last two digits. 624 = (6 x 100) + 24. 6 x 100 is divisible by 4, as is 24. To check the last 2 digits, if we keep subtracting 20's from the last 2 numbers, and get 4, 8, 12, 16, or 20, then the number is divisible by 4. So 3464 is divisible by 4 because 64 – 20 – 20 – 20 = 4. 3478 is not divisible by 4, since 78 – 20 – 20 – 20 = 18, and 18 is not divisible by 4.

To find out if a number is divisible by 7, take the last digit, double it, and subtract it from the rest of the number. If the result is divisible by 7, the number is divisible by 7. If you don't know the new number's divisibility, you can apply the rule again. For example, if you had 203, you would double the last digit to get 6, and subtract that from 20 to get 14.

To find out if a number is divisible by 8, check the last three digits. If the first digit of the last 3 digits is even, the number is divisible by 8 if the last two digits are divisible by 8. If the first digit of the last 3 digits is odd, subtract 4 from the last two digits; the number will be divisible by 8 if the resulting last two digits are. 23,888 is divisible by 8: The digit in the hundreds place is an even number, and 88 is divisible by 8. 23,886 is not divisible by 8: The digit in the hundreds place is an even number, but the 86 is not divisible by 8. 23,728 is divisible by 8: The digit in the hundreds place is odd, and 28 – 4 is divisible by 8.

(1) Find multiples of given numbers

Discussion

Concept page 32

Point out that the product of 3 and any number is a multiple of 3. Make sure your student sees the link between factors and multiples. If 12 is a multiple of 3, then 3 is a factor of 12. Ask her to find some other multiples of 3. Give her a number that is not a multiple of 3, such as 32, and ask her if it is a multiple of 3. We can determine if a number is a multiple of 3 by seeing if it is divisible by 3.

Tasks 1-8, pp. 33-34

1: We can find out if a number is a multiple of another in the same way that we would find out if a number is a factor of another by using division. Seeing if 3 divides into 36 without a remainder answers both of these questions.

4: This shows that we multiply by 1, 2, 3, and so on to find multiples. The first multiple of a number is the number itself.

6: Ask: Each number in the pattern is a multiple of what number? In (a), each is a multiple of 4, and in (b), each is a multiple of 6. Each number in the pattern is the first number multiplied by 1, 2, 3, and so on.

8: Give your student a hundred-chart and ask him to circle all the multiples of 3. Pick a circled number and have him add its digits together. Repeat with some other circled numbers. For each circles number, the sum of the digits will be a multiple of 3. If a number is a multiple of 3, then we know that it can be evenly divided by 3. Point to the number 99. Ask if it is a multiple of 3. It is. 9 + 9 = 18, which is a multiple of 3. Then ask him to add the two digits of the sum together. 1 + 8 = 9. That, too, is a multiple of 3. When testing if a number is a multiple of 3 by adding the digits, if the resulting number has more than one digit, we can continue to add the two digits of the sum together until we get a sum with one digit. The final sum will be 3, 6, or 9.

1. (a) yes
   (b) yes
2. (a) no
   (b) no
3. (a) yes
   (b) yes
   (c) yes
   (d) no
   (e) yes
4. 5, 10,15, 20
5. 9, 18, 27, 36
6. (a) 20, 24, 28
   (b) 30, 36, 42
7. (a) 4, 6, 8
   (b) 0, 5
8. yes

Activity

Give your student a hundred-chart and ask her to circle all the multiples of 6 and cross out the multiples of 8. Ask her what she can tell you about numbers that are both circled and crossed out. They are multiples of both 6 and 8. Tell her they are called common multiples of 6 and 8. Ask her to list the common multiples of 6 and 8. (24, 48, 72, 96) Tell her that 24 is the lowest common multiple of 8. See if she can see a pattern with the common multiples of 6 and

1	2	3	4	5	6	7	8	9	10
11	12	13	14	15	16	17	18	19	20
21	22	23	24	25	26	27	28	29	30
31	32	33	34	35	36	37	38	39	40
41	42	43	44	45	46	47	48	49	50
51	52	53	54	55	56	57	58	59	60
61	62	63	64	65	66	67	68	69	70
71	72	73	74	75	76	77	78	79	80
81	82	83	84	85	86	87	88	89	90
91	92	93	94	95	96	97	98	99	100

Common multiples of 6 and 8:
24, 48, 72, 96

8. They are all multiples of the lowest common multiple. See if she can think of an easy way to find one common multiple of 6 and 8. She can simply multiply 6 x 8 to get a common multiple of both 6 and 8.

Refer to the chart in the textbook on p. 34. Ask your student to use it to tell you which of the numbers are common multiples of 2 and 5.

Practice

Tasks 9-10, p. 35

9. 24, 36

10. (a) 3, 9
(b) 18, 36

Discussion

Task 11, p. 35

Point out that the girl is listing the multiples of the larger number, 5, and then seeing which one is also a multiple of 3. Lead your student to see that if we can recognize the multiples of one of the numbers, we don't have to list multiples of both to find common multiples. If we are just going to list the multiples of one of the numbers, we should use the larger number. See if your student can tell you why. The larger number has fewer multiples within 100, so we would be checking fewer numbers to see if they are the multiple of the smaller number.

Ask your student to use this method to find the first two common multiples of 9 and 6. (18, 36)

Workbook

Exercise 9, pp. 26-27 (answers p. 36)

Reinforcement

Extra Practice, Unit 1, Exercise 4, pp. 15-16

Ask your student the following "What number am I?" riddles:

⇒ I am less than 60. I am a common multiple of 10 and 15. (30)

⇒ I am greater than 10. I am a common factor of 28 and 42. (14)

⇒ I am a common multiple of 5 and 7. I am less than 100. I am not 35. (70)

⇒ I am an even number less than 79. 5, 4, and 3 are all factors of me. I am a multiple of 10. (60)

Game

Material: Hundred-chart, number cube labeled 1-6, markers for each player.

Procedure: Players take turns rolling the number cube and placing their markers on an unoccupied square on the hundred-chart with a number that is a multiple of the number rolled. (If a player rolls a 1, she can put her marker on any number.) The first person to get 3 in a row wins.

Alternative game: Players can place their marker on a space already occupied, replacing the marker that was there. The first player to get 5 in a row wins.

(2) Practice

Activity

Give your student a hundred-chart and ask him to circle the multiples of 3, and then use a different color to circle every other one, which are multiples of 6. Ask him for any pattern he might see. The multiples of 6 are all multiples of 3, and are all even. So if a number is a multiple of 3 and even, it has a factor of 6. (It has both 2 and 3 as factors.)

1	2	3	4	5	6	7	8	9	10
11	12	13	14	15	16	17	18	19	20
21	22	23	24	25	26	27	28	29	30
31	32	33	34	35	36	37	38	39	40
41	42	43	44	45	46	47	48	49	50
51	52	53	54	55	56	57	58	59	60
61	62	63	64	65	66	67	68	69	70
71	72	73	74	75	76	77	78	79	80
81	82	83	84	85	86	87	88	89	90
91	92	93	94	95	96	97	98	99	100

Start with a fresh hundred-chart and ask your student to circle the multiples of 9, or she can simply list the multiples of 9 within 100: 9, 18, 27, 36, 45, 54, 63, 72, 81, 90, and 99. Then ask her if she sees a pattern similar to the multiples of 3. The sum of the circled digits is always 9.

Summarize the common divisibility rules, which are applicable to all whole numbers.

A number is divisible by

⇒ 2 if the ones digit is 0, 2, 4, 6, or 8
⇒ 3 if the sum of the digits of the number is divisible by 3
⇒ 5 if the ones digit is 0 or 5
⇒ 6 if the number is divisible by both 2 and 3
⇒ 9 if the sum of the digits of the number is divisible by 9
⇒ 10 if the ones digit is 0

Start with a fresh hundred-chart and ask your student to circle the multiples of 4. Point out that the pattern repeats itself every two rows. Lead him to see that if we jump up two rows from a multiple of 4, we land on a multiple of 4. So if we can subtract 20's from a number and get a number that is divisible by 4, then the original number is divisible by 4.

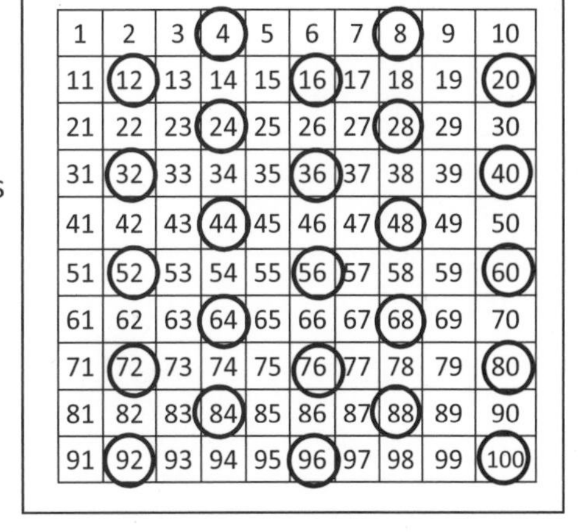

You can have your student experiment with larger numbers (and practice division) to see whether this holds true for larger numbers.

Practice

Practice C, pp. 36-37

Enrichment

Ask your student to find common multiples, or common factors, of more than 2 numbers. For example, ask her to find the smallest number that is a common multiple of 3, 5, and 9.

Tests

Tests, Unit 1, 4A and 4B, pp. 15-17

1. 1, 2, 3, 6, 9, 18

2. (a) 12, 15
   (b) 25, 30, 35, 40

3. (a) 8 (b) 9
   (c) 7 (d) 3
   (e) 4 (f) 3

4. (a) 1, 2, 4, 8
   (b) 1, 3, 5, 15
   (c) 1, 2, 4, 5, 10, 20
   (d) 1, 2, 5, 10, 25, 50
   (e) 1, 3, 5, 15, 25, 75
   (f) 1, 2, 7, 14, 49, 98

5. (a) 3 (b) 2 or 4 (c) 3

6. (a) 2, 4, 6, 8
   (b) 6, 12, 18, 24
   (c) 8, 16, 24, 32

7. (a) 12 and multiples of 12
   (b) 20 and multiples of 20
   (c) 12 and multiples of 12

8. 2, 3, 5, 7

Workbook

Exercise 6, pp. 19-20

1. (a) 300 (b) 1320

2. (a) 6000 (b) 36,300

3. (a) 46,000 (b) 236,000

4. (a) 245,000
 (b) 248,000

5. (a) 43,190
 (b) 14,600
 (c) 83,000
 (d) 2,000,000
 (e) 20,000,000

6. (a) $440,000 (b) $530,000
 (c) $2,610,000 (d) $3,970,000
 (e) $5,990,000 (f) $6,230,000

7. (a) $3,100,000 (b) $5,700,000
 (c) $18,300,000 (d) $25,000,000
 (e) $43,800,000 (f) $48,900,000
 (g) $328,500,000 (h) $693,500,000

Exercise 7, pp. 21-22

1. 1, 2, 4, 5, 10, and 20

2. (a) 2 and 6
 (b) 1 and 8
 (c) 3 and 7

3. (a) 4 (b) 3
 (c) 9 (d) 8
 (e) 8 (f) 8
 (g) 9 (h) 9
 (i) 10 (j) 8

4. (a) 8
 4
 1, 2, 4, and 8

 (b) 15
 5
 1, 3, 5, and 15

5. 13, 19

Exercise 8, pp. 23-25

1. (a) no (b) yes

2.

Number	Is 3 a factor?	Is 4 a factor?	Is 5 a factor?
30	yes	no	yes
36	yes	yes	no
48	yes	yes	no
60	yes	yes	yes
75	yes	no	yes
84	yes	yes	no

3. (a) yes
 (b) no
 (c) yes

4. (a) 64 = **8** x **8**
 1, 2, 4, 8, 16, 32, 64
 (b) 1, 2, 3, 4, 6, 8, 9, 12, 18, 24, 36, 72
 (c) 1, 2, 3, 4, 6, 7, 12, 14, 21, 28, 42, 84

5. (a) 64 = 2 x 8 x **2** x **2**
 (b) 84 = 6 x **2** x **7**
 (c) 45 = **3** x **3** x 5
 (d) 72 = **2** x 4 x **3** x **3**

6. (a) 4 (b) 3
 (c) 5 (d) 10
 (e) 16 (f) 8

Exercise 9, pp. 26-27

1. (a) 6; 12; 18; 24
 (b) 7, 14, 21, 28, and 35

2. 6, 8, 10, 12, 14
 9, 12, 15, 18, 21
 12, 16, 20, 24, 18
 18, 24, 30, 36, 42
 24, 32, 40, 48, 56
 30, 40, 50, 60, 70

3. (a) 6, 12
 (b) 8, 16
 (c) 9, 18, 27, 36
 6, 12, 18, 24, 30, 36
 18, 36
 (d) 8, 16, 24, 32, 40, 48
 6, 12, 18, 24, 30, 36, 42, 48
 24, 48

Chapter 5 — Order of Operations

Objectives

- Learn and use the rules for order of operations.
- Apply order of operations to expressions with parentheses.
- Write expressions for two-step word problems that show both steps in a single expression.
- Understand that performing the same operation on both sides of an equation does not change the equality.

Vocabulary

- Expression
- Equation

Material

- Number cards 0-9, 4 sets
- Operation cards, +, –, x, and ÷, 8 sets

Notes

Students learned in earlier levels of *Primary Mathematics* that the order in which they add or multiply two numbers does not affect the result (the commutative properties of addition and multiplication). They also learned that when they add or multiply any three numbers, which two numbers they add or multiply together first does not affect the final result (the associative properties of addition and multiplication). Changing the order of the numbers can simplify the problem. Subtraction and division are not commutative or associative (unless we rewrite the problems as addition of negative or multiplication of the inverse, neither of which has been taught yet). They must be done in order from left to right.

In this chapter, your student will learn that if an expression involves all four operations, by convention, she first does multiplication and division from left to right, and then addition and subtraction from left to right. Multiplication and division take precedence over addition and subtraction. If an expression includes parentheses, she does the operations within the parentheses first, following order of operations for the expression within the parentheses.

Do <u>not</u> teach your student a mnemonic such as "Please Excuse My Dear Aunt Sally" for Parentheses, Exponents, Multiplication and Division, Addition and Subtraction. This memory device often leads to confusion later when students simply rely on the mnemonic and forget that multiplication and division are treated at the same level as are addition and subtraction. The confusion resulting from this mnemonic is quite prevalent.

$\overbrace{8 + 3 + 2}^{10} + 4 + 7 = 10 + 10 + 4 = 24$ (8 + 2 = 10; 3 + 7 = 10)
$\overbrace{5 \times 7 \times 4}^{20} = 20 \times 7 = 140$ (5 x 4 = 20)
$\underline{10 - 4} - 3 = 6 - 3 = 3$ (**not** $10 - \underline{4 - 3} = 10 - 1 = 9$)
$\underline{32 \div 4} \div 2 = 8 \div 2 = 4$ (**not** $32 \div \underline{4 \div 2} = 32 \div 2 = 16$)
$10 - \underline{4 \div 2} \times 5 + 3$ $= 10 - \underline{2 \times 5} + 3$ $= \underline{10 - 10} + 3$ $= \underline{0 + 3}$ $= 3$
$23 - (8 + \underline{2 \times 5}) \div 6$ $= 23 - (\underline{8 + 10}) \div 6$ $= 23 - \underline{18 \div 6}$ $= \underline{23 - 3}$ $= 20$

(1) Solve expressions with mixed operations

Activity

Write the following expressions, ask your student to solve them, and then discuss them.

⇒ $8 + 3 + 2 + 4 + 7$

Since we can add in any order we can often make the problem simpler by changing the order.

$8+3+2+4+7$
$= \underline{8+2}+\underline{3+7}+4$
$= 10 + 10 + 4$
$= 24$

⇒ $5 \times 7 \times 4$

Since we can also multiply in any order, changing the order can sometimes simplify the problem.

$5 \times 7 \times 4$
$= 5 \times 4 \times 7$
$= 20 \times 7$
$= 140$

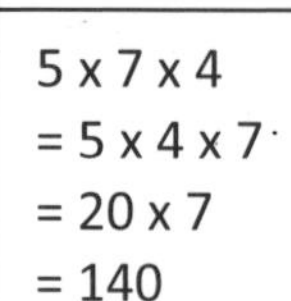

⇒ $10 - 4 - 3$

Ask your student to first solve the problem by subtracting in order from left to right and then solve it in a different order. The answers are different. We cannot subtract in any order.

$\underline{10 - 4} - 3$
$= 6 - 3$
$= 3$ ✓

$10 - \underline{4 - 3}$
$= 10 - 1$
$= 9$ ✗

Ask for suggestions as to why subtraction is different from addition and cannot be done in any order. In addition problems, we are simply putting parts together and can do so in any order. In subtraction we are subtracting from a whole. Both 4 and 3 must be subtracted from the whole. If we subtract 3 from 4 and then subtract that answer from 10, 4 becomes a whole and we end up only subtracting the difference between 4 and 3 from 10. You can ask your student to try to diagram both situations with number bonds.

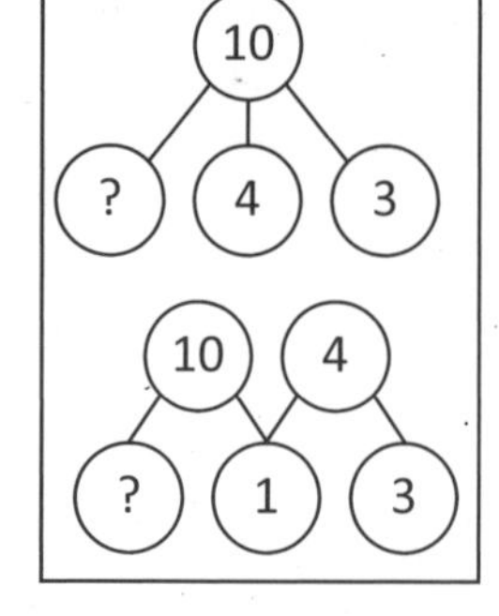

Optional: If the "minus" operation is kept with the number, subtraction can be done in any order. If you feel your student is capable of keeping the number always with the operation, you can show this as follows. Use 10 multilink cubes linked together. Have him illustrate the process of $10 - 4 - 3$. Get him to realize that he gets the same answer when he takes 3 away first. Tell him this works because the minus sign is kept with the number being subtracted. If he does that, he can change the order in which we *subtract* the two numbers from 10. Both are then kept as parts that he takes away from the whole. He cannot, however, subtract 3 from 4 first, and then subtract that answer from 10. If he does that, 4 becomes a whole, not a part.

⇒ $32 \div 4 \div 2$

Ask your student to solve this by dividing first from left to right and then try a different order. Division also involves starting out with a whole, so we cannot do it in any order.

$\underline{32 \div 4} \div 2$
$= 8 \div 2$
$= 4$ ✓

$32 \div \underline{4 \div 2}$
$= 32 \div 2$
$= 16$ ✗

Optional: As with subtraction, we can first divide 32 by 2 and then by 4 and get the same answer. That is not the same as dividing 4 by 2 and then dividing 32 by that quotient.

⇒ 25 – 10 + 5

When a problem has both addition and subtraction, it is done in order from left to right. Your student can try a different order to see that the answers are not the same.

25 – 10 + 5
= 15 + 5
= 20

Optional: Again, lead your student to see that we can change the order if we keep the operation with the number. We can add 5 to 25, and then subtract 10 and get the same answer. But we cannot add 5 to 10, and then subtract that sum from 25 to get the same answer.

⇒ 30 ÷ 6 x 2

Tell your student that when a problem has both multiplication and division, we also do it in order from left to right.

30 ÷ 6 x 2
= 5 x 2
= 10

Practice

Tasks 1-2, p. 39

1.	(a) 10	(b) 24	(c) 23
	(d) 32	(e) 147	(f) 99
	(g) 29	(h) 31	(i) 75
2.	(a) 64	(b) 5	(c) 27
	(d) 120	(e) 3	(f) 10
	(g) 160	(h) 1	(i) 8

Discussion

Concept page 38

Ask your student what the answer would be if we were to add first to solve the 10 + 4 x 3. The answer would be 42 and would not give the total number of stamps. Mathematicians have agreed on a specific order of operations for expressions that have more than one operation (+, –, x, or ÷). We first do all the multiplication and division in order from left to right, and then all the addition and subtraction in order from left to right.

Practice

Task 3, p. 39

Workbook

Exercise 10, pp. 28-29 (answers p. 47)

(Note: In the 2008 printing of the workbook, problem 3(b) has an error. It should read **60 ÷ 5 – 6 x 2**.)

Game

Material: 4 sets of number cards 0-9, 8 sets of operation cards **+**, -, x, and ÷.

Procedure: For each round, deal 6 cards to each player. Turn one more card face up and place in the center. Place the operation cards in the center so they are available to all players. Each player must form an expression whose answer is the number in the middle, using any of the four operations. For example, if the card in the middle is a 6, a player with the cards 3, 4, 6 and 9 can make the expression 6 x 3 ÷ 9 + 4. The player gets one point for each number card used. The first player who gets 25 points wins.

3. (a) 9 + 3 x 6
= 9 + 18
= **27**

(b) 27 – 12 ÷ 3
= 27 – 4
= **23**

(c) 4 + 5 x 8
= 4 + 40
= **44**

(d) 80 – 5 x 10
= 80 – 50
= **30**

(e) 54 – 48 ÷ 6
= 54 – 8
= **46**

(f) 9 + 81 ÷ 9
= 9 + 9
= **18**

(g) 56 – 8 x 5 + 4
= 56 – 40 + 4
= 16 + 4
= **20**

(h) 70 + 40 ÷ 5 x 4
= 70 + 8 x 4
= 70 + 32
= **102**

(i) 96 ÷ 8 – 6 x 2
= 12 – 6 x 2
= 12 – 12
= **0**

(j) 6 + 54 ÷ 9 x 2
= 6 + 6 x 2
= 6 + 12
= **18**

(k) 49 – 45 ÷ 5 x 3
= 49 – 9 x 3
= 49 – 27
= **22**

(l). 62 + 42 ÷ 7 – 6
= 62 + 6 – 6
= 68 – 6
= **62**

(2) Solve expressions with mixed operations and parentheses

Activity

Refer to textbook p. 38 again. Tell your student that the boy put stamps on 3 more pages. On each of them, he put 10 of one kind and 4 of another. Ask her how we could write one expression to find how many stamps there are on the 3 pages. Lead her to see that we can easily find the answer in two steps: 10 + 4 = 14, then 14 x 3 = 42. There are 42 stamps. But if we write a single expression, 10 + 4 x 3, we would need a way to show that the addition needs to be done first now. Write the equation with parenthesis. Tell her that if the expression has parentheses, we treat the problem in parentheses as an expression and solve it first.

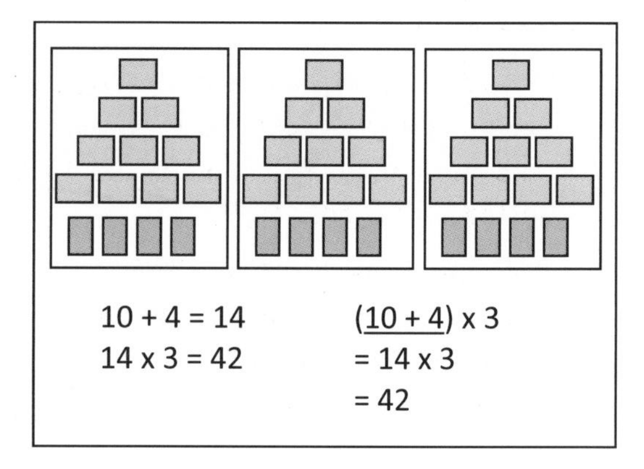

Discussion

Task 4, p. 39

4. 11

Practice

Tasks 5-6, pp. 39-40

5. (a) 9 + (36 + 16)
 = 9 + 52
 = **61**

 (b) 100 – (87 – 13)
 = 100 – 74
 = **26**

 (c) 99 – (87 + 12)
 = 99 – 99
 = **0**

 (d) 18 x (5 x 2)
 = 18 x 10
 = **180**

 (e) 49 ÷ (7 x 7)
 = 49 ÷ 49
 = **1**

 (f) 100 x (27 ÷ 9)
 = 100 x 3
 = **300**

6. (a) 60 ÷ (4 + 8)
 = 60 ÷ 12
 = **5**

 (b) 20 – 2 x (18 ÷ 6)
 = 20 – 2 x 3
 = 20 – 6
 = **14**

 (c) 25 + (5 + 7) ÷ 3
 = 25 + 12 ÷ 3
 = 25 + 4
 = **29**

 (d) (22 + 10) ÷ 8 x 5
 = 32 ÷ 8 x 5
 = 4 x 5
 = **20**

 (e) (50 – 42) ÷ 2 x 7
 = 8 ÷ 2 x 7
 = 4 x 7
 = **28**

 (f) 100 ÷ 10 x (4 + 6)
 = 100 ÷ 10 x 10
 = 10 x 10
 = **100**

Workbook

Exercise 11, pp. 30-31 (answers p. 47)

Enrichment

Ask your student to use the digit 4 four times in an expression with any of the four operations, with or without parentheses, to make up the numbers from 0 to 9. More than one solution is possible for some numbers. Here are some possible solutions:

(4 + 4) – (4 + 4) = 0
(4 + 4) ÷ (4 + 4) = 1
4 ÷ 4 + 4 ÷ 4 = 2
(4 + 4 + 4) ÷ 4 = 3
4 x (4 – 4) + 4 = 4
(4 x 4 + 4) ÷ 4 = 5
(4 + 4) ÷ 4 + 4 = 6
4 + 4 – 4 ÷ 4 = 7
4 + 4 + 4 – 4 = 8
4 ÷ 4 + 4 + 4 = 9

(3) Write expressions with mixed operations

Discussion

Task 7, p. 40

Both (a) and (b) can be solved in two steps. We can also write an equation that shows both steps together by using parentheses and putting the expression for the first step in the parentheses instead of the answer.

Tell your student that he is now required to write both steps in a single expression unless he is requested to do differently in the directions for the problem.

Note: Your student will see the benefits of expressions with parentheses later when learning about math formulas.

7. (a) \$20 – (\$8 + \$5)
= \$20 – \$13
= **\$7**
She had \$7 left.
(b) (\$20 – \$8) + \$5
= \$12 + \$5
= **\$17**
He has \$17 now.
(c) no

Practice

Task 8, p. 40

8. (a) (5 x 8) + (2 x 16)
(b) = 40 + 32 = **72**
She bought 72 fruits.

Discussion

Tasks 9-10, p. 41

Have your student read the definition of an equation at the top of p. 41 and compare that with the definition of an expression on p. 39.

For Task 9, if you change both sides of an equation in the same way, the equation remains true. Be sure your student understands that adding 10 is the same as adding 2 x 5.

9. (a) 18 = 18 yes
(b) 18 = 18 yes
(c) 80 = 80 yes
(d) 80 = 80 yes
10. (a) 24 (b) 9
(c) 100 (d) 12

Write an equality, such as 10 = 10, and let your student add to it by adding or multiplying the same number or an expression that evaluates to the same number to both sides.

For Task 10, your student should be able to determine the missing number without completely solving the expressions on the left-hand side.

Workbook

Exercise 12, pp. 32-33 (answers p. 48)

Reinforcement

Extra Practice, Unit 1, Exercise 5, pp. 17-18

Test (and Enrichment)

Tests, Unit 1, 5A and 5B, pp. 19-22

The tests for this chapter include some problems that are challenging, which might be better suited as enrichment problems rather than test problems. They are problems 4-5 in the A test, and problems 6-10 in the B test.

Chapter 6 – Negative Numbers

Objectives

- Use negative numbers in practical situations.
- Locate negative numbers on a number line.
- Order and compare integers.
- Count on or back from a given integer.

Vocabulary

- Positive numbers
- Negative numbers

Positive numbers are numbers greater than 0. For example, 1, 10, 18, and 456 are **positive numbers**. For each positive number, there is a **negative number** that is its opposite. Negative numbers are written with a negative sign in front of them. For example, the opposites of 1, 10, 18, and 456 are –1, –10, –18, and –456. Even though the negative sign looks just like a minus sign, the two symbols, the negative sign and the numeral, together represent a *single* number. So –1 represents the number negative 1, and the sign before the 1 does not refer to subtraction. Later, in algebra, students will make the connection between negative numbers and subtraction.

At this level, your student will deal only with negative whole numbers. All whole numbers, including negative numbers, are called integers.

In this chapter, your student is introduced to negative numbers first in context of practical situations where measurements less than 0 make sense, such as temperatures below 0 degrees, distances below sea level or ground level, the time prior to a specific event, like a shuttle blast-off, or owing money.

Negative numbers will then be related to a simple extension of the number line in a direction less than 0. Negative numbers are less than 0 and are represented on a horizontal number line to the left of 0 and on a vertical number line below 0.

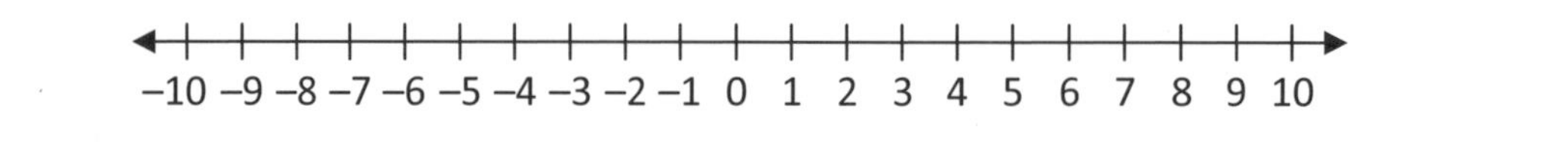

Zero is neither positive nor negative.

As with positive integers, negative integers have order. On the number line, larger numbers are to the right (or above for vertical number lines) of smaller numbers. 10 is larger than, or greater than, 1. –10, however, is smaller than –1, since –1 is to the right of –10. This can be confusing for your student, who might focus on the numerical part of the negative number. Initially, he will use a number line to order integers. With larger numbers, he can imagine where the numbers will be on a number line. –999 is greater than –1000 because it is to the right of –1000 on the number line.

(1) Understand the concept of negative numbers

Activity

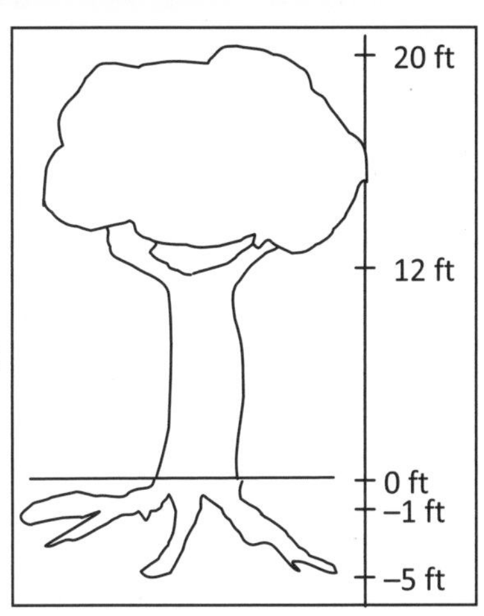

Draw a simple tree or some other type of plant, showing the ground level, the trunk and branches above the ground level, and the roots below. Draw a vertical number line next to it. Discuss the height of the tree and mark the different heights on the number line, with 0 at ground level. For example, the tree is 20 feet tall, and a branch comes out at 12 feet. Tell your student that when we say the tree is 20 ft tall, that is the distance up from the ground. So ground level is 0 ft. Then discuss the depth of the roots, again marking the values on the number line, going down from 0. For example, the roots go to a depth of 5 feet. A large root goes just under the ground at 1 foot.

Tell your student that we now have distances both above and below the ground, each relative to ground level. We need to distinguish between the two. We can think of the distances below ground level as less than 0, and call them negative numbers. We mark them with a "–" sign just in front of the number. We read such a number using the word negative: *negative 5* or *negative 1*. The negative number looks like a minus sign, but it is part of the number.

Discuss other ways negative numbers are used.

Discussion

Concept p. 42

Tasks 1-3, p. 43

Point out that on the number line on p. 43, the numbers after 0 are marked with a positive sign, which looks like a plus sign. A number without a sign is always positive. Usually only negative numbers are always marked with a sign. Tell your student that usually when we draw number lines going from left to right (horizontally), negative numbers are always to the left of 0. When we draw them going up and down (vertically) as with the number line on p. 42 showing the height above and below sea level, negative numbers are always below 0.

9000 m
3400 m
200 m
5000 m
5000 m

1. –30 degrees (negative thirty degrees)
2. \$–10, or –\$10 (negative 10 dollars)
3. –10 (negative ten)

Practice

Tasks 4-6, p. 44

4. A: –8
   B: –4
   C: –1
   D: 3
5. –25; –15; 5; 10; 20
6. (a) –10, –9, –8, –7, –6, –5, –4, –3, –2, –1, 0, 1, 2, 3, 4, 5, 6, 7, 8, 9, 10
   (b) –10, –8, –6, –4, –2, 0, 2, 4, 6, 8, 10

Workbook

Exercise 13, pp. 34-35 (answers p. 48)

Reinforcement

Ask your student to count by other multiples, such as 3, starting from a negative number, such as –15.

(2) Compare and order positive and negative numbers

Discussion

Task 7, p. 44

7. (a) −5 is greater than −10.
   (b) −5 > −10

Have your student point to each pair of numbers on the number line for each comparison. You may want to draw a larger number line. Make sure he understands that with a negative number such as −5, the negative sign and the number together make the number, not the 5 by itself. The numerical portion alone simply tells us how far the number is from 0. −5 is a different number than 5. Numbers are arranged in increasing order from left to right on the number line, so every number is less than any number on its right and greater than any number on its left. The farther a negative number is from 0, the larger will be the numerical portion of the written number, but the smaller the actual number will be.

Practice

Tasks 8-11, p. 45

8. (a) −7 < 4 (b) 3 > −9
   (c) 8 > 2 (d) −8 < −2
   (e) −10 > −15 (f) −99 > −100
9. −8, −5, 6, 7, 10
10. (a) 6
    (b) −4
    (c) −1
    (d) −11
    (e) 0
11. (a) 2, 1, 0, −1, −2
    (b) 0, −2, −4, −6

Workbook

Exercise 14, pp. 36-37 (answers p. 48)

Reinforcement

Have your student find other numbers that are 1, 2, or 3 more than or less than a given negative number. You can extend this to larger numbers and place-value concepts. For example, ask her to find the number that is 2000 less than −34,590. (−36,590)

Extra Practice, Unit 1, Exercise 6, pp. 19-20

(3) Practice

Practice

Practice D, pp. 46-47

In problem 2, your student should be able to find the missing number without evaluating the left side for all except (e). In (e), she will have to evaluate the left-hand side completely.

In problem 3, your student should find the answer without having to evaluate any of the expressions.

Tests

Tests, Unit 1, 6A and 6B, pp. 23-26

1. (a) $\underline{9 \times 7} - \underline{6 \div 6}$
   $= 63 - 1$
   $= \mathbf{62}$

   (b) $8 - \underline{12 \div 4} + 5$
   $= \underline{8 - 3} + 5$
   $= 5 + 5$
   $= \mathbf{10}$

   (c) $8 + 1 - \underline{2 \times 0}$
   $= 8 + 1 - 0$
   $= \mathbf{9}$

   (d) $12 \times (\underline{7 - 6}) \div 6$
   $= \underline{12 \times 1} \div 6$
   $= 12 \div 6$
   $= \mathbf{2}$

   (e) $(\underline{12 - 8}) \div 4 + 5$
   $= \underline{4 \div 4} + 5$
   $= 1 + 5$
   $= \mathbf{6}$

   (f) $(\underline{8 + 2}) \times 10 \div 2$
   $= \underline{10 \times 10} \div 2$
   $= 100 \div 2$
   $= \mathbf{50}$

   (g) $\underline{3 \times 7} + \underline{3 \div 3}$
   $= 21 + 1$
   $= \mathbf{22}$

   (h) $1 + 2 \times (\underline{4 - 3})$
   $= 1 + \underline{2 \times 1}$
   $= 1 + 2$
   $= \mathbf{3}$

2. (a) 12
   (b) 3
   (c) 7
   (d) 4
   (e) 35
   (f) 20

3. (c) 24 + (2 x 6)

4. A: –10; B: –6; C: –2; D: 6

5. (a) –9, –7, 6, 8, 10
   (b) –2, –1, 0, 1, 2

Review 1

Review

Review 1, pp. 48-50

1(d): This problem is not meant to be how we would read the resulting number. If your student has trouble with this problem, have her first write twelve hundred, 1200, then twelve hundred thousand, 1200,000

1(e): You can have your student write the number for seventy thousand, 70,000, and then add six 0's for seventy thousand millions.

14(a-f): Your student may start out solving the problems by trying out different numbers. You can point out that he can solve the problems in the same way as he would solve a problem involving division of 2-digit numbers with a remainder. For example, to find the greatest whole number that can be put in the blank for 3 x ___ < 20, we can think of the answer to 20 ÷ 3, which is 6 with a remainder. 6 is therefore the greatest number that can be put into the blank.

Workbook

Review 1, pp. 38-39 (answers p. 48)

Problem 6 in Review 1 can be solved by simply seeing if 6 divides evenly into 4598. We can also apply rules of divisibility.

Tests

Tests, Unit 1 Cumulative Tests A and B, pp. 27-32

1. (a) 3730 (b) 27,089
   (c) 100,000 (d) 1,204,803
   (e) 70,000,000,000 (f) –48
2. (a) fifteen thousand, seven hundred eighty
   (b) five million, three hundred six thousand, nine hundred three
   (c) negative twenty thousand, four
3. (a) 600,000 (b) 600 (c) 60,000
4. (a) 5980; 6080 (b) 34,465; 35,465
   (c) 183,700; 193,700 (d) 0; –10; –20
5. (a) > (b) < (c) > (d) =
6. (a) 30,016,000; 30,061,000; 30,160,000; 30,601,000
   (b) 20,990; 29,909; 29,999; 90,000
   (c) –21, –8, –2, 3, 11, 13
   (d) –18, –12, –6, 2, 9, 15, 21
7. 6200
8. 29,000 ft
9. (a) 14,000,000 (b) 13,900,000
   (c) 13,940,000 (d) 13,940,000
10. (a) No. 2 + 8 = 10, which is not divisible by 3.
    (b) Yes. 60 has 0 in the ones place.
    (c) No. 80 is even, but 8 is not divisible by 3.
    (d) No. There is a 2 in the ones place.
    (e) Yes. 100 is a multiple of 20 which is a multiple of 4. Or, 100 ÷ 4 = 25.
11. 1, 31
12. (a) 35 x 3 = 5 x **7** x 3
    (b) 25 x 24 = 25 x 4 x **6**
    (c) 4 x 24 = 4 x 8 x 3 = 8 x 4 x 3 = 8 x **12**
    (d) 64 x 2 = 32 x 2 x 2 = 32 x **4**
13. (a) $4 + \underline{32 \div 8}$
    $= 4 + 4$
    $= \mathbf{8}$
    (b) $6 \times (\underline{22 - 12})$
    $= 6 \times 10$
    $= \mathbf{60}$
    (c) $40 - (\underline{3 \times 12}) \div 6$
    $= 40 - \underline{36 \div 6}$
    $= 40 - 6$
    $= \mathbf{34}$
    (d) $\underline{60 \div 10} - (\underline{4 + 2})$
    $= 6 - 6$
    $= \mathbf{0}$
14. (a) 6 (b) 6
    (c) 8 (d) 7
    (e) 8 (f) 4
    (g) 14 (h) 27

Workbook

Exercise 10, pp. 28-29

1. (a) $\underline{8 + 12} + 20$
$= \underline{20 + 20}$
$= \mathbf{40}$

   (b) $\underline{40 - 14} - 9$
$= \underline{26 - 9}$
$= \mathbf{17}$

   (c) $\underline{26 + 8} - 9$
$= \underline{34 - 9}$
$= \mathbf{25}$

   (d) $\underline{21 - 5} + 8$
$= \underline{16 + 8}$
$= \mathbf{24}$

   (e) $\underline{3 \times 5} \times 8$
$= \underline{15 \times 8}$
$= \mathbf{120}$

   (f) $\underline{36 \div 3} \div 4$
$= \underline{12 \div 4}$
$= \mathbf{3}$

   (g) $\underline{4 \times 9} - 3$
$= \underline{36 - 3}$
$= \mathbf{33}$

   (h) $\underline{64 \div 8} \times 5$
$= \underline{8 \times 5}$
$= \mathbf{40}$

2. (a) $\underline{24 \div 6} \times 8$
$= \underline{4 \times 8}$
$= \mathbf{32}$

   (b) $140 - \underline{40 \times 3}$
$= \underline{140 - 120}$
$= \mathbf{20}$

   (c) $46 + \underline{32 \div 8}$
$= \underline{46 + 4}$
$= \mathbf{50}$

   (d) $100 - \underline{60 \div 4}$
$= \underline{100 - 15}$
$= \mathbf{85}$

   (e) $\underline{8 \times 6} + 14$
$= \underline{48 + 14}$
$= \mathbf{62}$

   (f) $12 + \underline{18 \div 6}$
$= \underline{12 + 3}$
$= \mathbf{15}$

   (g) $\underline{6 \times 10} - 5$
$= \underline{60 - 5}$
$= \mathbf{55}$

   (h) $72 + \underline{6 \times 6}$
$= \underline{72 + 36}$
$= \mathbf{108}$

3. (a) $20 + \underline{4 \div 4} - 4$
$= \underline{20 + 1} - 4$
$= \underline{21 - 4}$
$= \mathbf{17}$

   (b) $\underline{60 \div 5} - \underline{6 \times 2}$
$= 12 - 12$
$= \mathbf{0}$
* (error in 2008 printing)

   (c) $80 - \underline{36 \div 4} \times 3$
$= 80 - \underline{9 \times 3}$
$= \underline{80 - 27}$
$= \mathbf{53}$

   (d) $32 + 8 + \underline{10 \times 2}$
$= \underline{32 + 8} + 20$
$= \underline{40 + 20}$
$= \mathbf{60}$

   (e) $52 - \underline{35 \div 7} - \underline{7 \times 2}$
$= \underline{52 - 5} - 14$
$= \underline{47 - 14}$
$= \mathbf{33}$

   (f) $\underline{9 \times 8} - \underline{6 \times 10}$
$= \underline{72 - 60}$
$= \mathbf{12}$

   (g) $\underline{7 \times 8} + \underline{24 \div 8}$
$= \underline{56 + 3}$
$= \mathbf{59}$

   (h) $\underline{63 \div 9} + \underline{20 \div 10}$
$= \underline{7 + 2}$
$= \mathbf{9}$

Exercise 11, pp. 30-31

1. (a) $9 + (\underline{26 - 15})$
$= \underline{9 + 11}$
$= \mathbf{20}$

   (b) $90 - (\underline{4 + 6})$
$= \underline{90 - 10}$
$= \mathbf{80}$

   (c) $12 - (\underline{10 - 8})$
$= \underline{12 - 2}$
$= \mathbf{10}$

   (d) $(\underline{31 - 20}) - 8$
$= \underline{11 - 8}$
$= \mathbf{3}$

   (e) $8 \times (\underline{3 \times 2})$
$= \underline{8 \times 6}$
$= \mathbf{48}$

   (f) $20 \div (\underline{4 \div 2})$
$= \underline{20 \div 2}$
$= \mathbf{10}$

   (g) $9 \times (\underline{15 \div 3})$
$= \underline{9 \times 5}$
$= \mathbf{45}$

   (h) $15 \div (\underline{5 \times 3})$
$= \underline{15 \div 15}$
$= \mathbf{1}$

2. (a) $(\underline{9 + 6}) \div 5$
$= \underline{15 \div 5}$
$= \mathbf{3}$

   (b) $2 \times (\underline{9 - 4})$
$= \underline{2 \times 5}$
$= \mathbf{10}$

   (c) $12 \div (\underline{8 - 6})$
$= \underline{12 \div 2}$
$= \mathbf{6}$

   (d) $(\underline{4 + 6}) \times 5$
$= \underline{10 \times 5}$
$= \mathbf{50}$

   (e) $10 \times (\underline{15 \div 5})$
$= \underline{10 \times 3}$
$= \mathbf{30}$

   (f) $(\underline{51 - 44}) \div 7$
$= \underline{7 \div 7}$
$= \mathbf{1}$

   (g) $72 \div (\underline{9 - 3})$
$= \underline{72 \div 6}$
$= \mathbf{12}$

   (h) $(\underline{28 - 18}) \times 10$
$= \underline{10 \times 10}$
$= \mathbf{100}$

3. (a) $20 + (\underline{8 + 4}) \div 3$
$= 20 + \underline{12 \div 3}$
$= \underline{20 + 4}$
$= \mathbf{24}$

   (b) $16 + (\underline{9 - 3}) \times 5$
$= 16 + \underline{6 \times 5}$
$= \underline{16 + 30}$
$= \mathbf{46}$

   (c) $3 \times (\underline{4 + 2}) \div 2$
$= \underline{3 \times 6} \div 2$
$= \underline{18 \div 2}$
$= \mathbf{9}$

   (d) $7 \times (\underline{13 - 6}) - 19$
$= \underline{7 \times 7} - 19$
$= \underline{49 - 19}$
$= \mathbf{30}$

   (e) $60 + (\underline{18 + 7}) \div 5$
$= 60 + \underline{25 \div 5}$
$= \underline{60 + 5}$
$= \mathbf{65}$

   (f) $8 \times (\underline{11 - 8}) \div 6$
$= \underline{8 \times 3} \div 6$
$= \underline{24 \div 6}$
$= \mathbf{4}$

   (g) $\underline{24 \div 6} + 3 \times (\underline{6 - 4})$
$= 4 + \underline{3 \times 2}$
$= \underline{4 + 6}$
$= \mathbf{10}$

   (h) $10 + (\underline{28 - 8}) \div 5 \times 2$
$= 10 + \underline{20 \div 5} \times 2$
$= 10 + \underline{4 \times 2}$
$= \underline{10 + 8}$
$= \mathbf{18}$

Workbook

Exercise 12, pp. 32-33

1. (a) (50 – 32) + 18 = 36
   (b) (20 – 5) – 10 = 5
   (c) (15 x 2) + 10 = 40

2. (a) ✓ 24 = 24
   (b) × 99 ≠ 88
   (c) × 35 ≠ 15
   (d) ✓ 7 = 56 – 49
   (e) ✓ 1 x 5 = 5
   (f) × 2 x 3 = 6 but 5 x 2 ≠ 8
   (g) ✓ 4 + 2 = 8 – 2

3. (a) (2 + 4) ÷ 2 = 3
   (b) 6 – (2 x 3) = 0
   (c) (2 x 4) – 3 + 2 = 7
   (d) 2 x (4 – 3) + 2 = 4
   (e) 2 x 4 – (3 + 2) = 3

Exercise 13, pp. 34-35

1. –154 degrees Celsius
2. (a) $14
   (b) $–5 or –$5
   (c) $–12 or –$12
3. – 5, – 4, –3, –2, –1
   (b) to the left of 0
   (c) above 0
4. (a) –2
   (b) 2
   (c) 1
   (d) 0
   (e) –3
   (f) –2
5. –4; –2; 0; 6; 10

Exercise 14, pp. 36-37

1. (a) 4 > –3 (b) –6 < 6
   (c) –7 < 0 (d) –2 < –1
   (e) –99 < –98 (f) –20 < –10
   (g) 50 > –50 (h) 0 > –100

2. (a) –5 (b) –99
   (c) –10 (d) –4

3. (a) –8, –5, 0, 4 (b) –15, –1, 1, 2
   (c) –9, –7, 8, 10 (d) –100, –-99, 99, 100

4. (a) 10, 4, –5, –14
   (b) 1, 0, –2, –4,
   (c) 40, 20, –30, –50
   (d) –5, –7, –9, –12

5. (a) –8
   (b) –12
   (c) –2
   (d) –7
   (e) –2
   (f) –100

Review 1, pp. 38-39

1. (a) 14,060 (b) 26,008
   (c) 10,032,520 (d) 400,069
   (e) 59,000,400

2. (a) five hundred thousand, six
   500,000 + 6
   (b) thirty-four million, four thousand, one hundred twenty
   30,000,000 + 4,000,000 + 4000 + 100 + 20
   (c) twelve thousand, twenty-five
   10,000 + 2,000 + 20 + 5

3. (a) 7; 700,000
   (b) tens
   (c) 4; 2

4. (a) 5 x 24 = 5 x 6 x **4**
   (b) 2 x 180 = 2 x 2 x 90 = 4 x **90**
   (c) 16 + 4 = 20, so the missing number is **2**.
   (d) 48 ÷ (4 x 2) – 2 = 2 x ___
   48 ÷ 8 – 2
   6 – 2 = 2 x **2**

5. 453,500

6. 4 + 5 + 9 + 8 = 26, which is not divisible by 3. 6 is **not** a factor of 4598.

7. 1, 2, 3, 6, 9, 18, 27, 54

8. 29

9. 20, 40

10. (a) 140, 260, 296, 435, 463, 503, 540, 870
    (b) –22, –19, –17, –16, 18, 20, 21

Unit 2 – The Four Operations of Whole Numbers

Chapter 1 – Addition and Subtraction

Objectives

- Add and subtract multi-digit numbers using the standard algorithms.
- Find the missing number (subtrahend, minuend, or addend) in equations involving addition and subtraction.
- Use mental math strategies to subtract from 10, 100, and 1000.
- Use mental math strategies to add and subtract some multi-digit numbers.
- Use approximation to estimate the answer to addition and subtraction problems.
- Determine if a problem requires an estimated or an exact answer.
- Review modeling method for solving word problems involving addition and subtraction.
- Solve word problems involving addition and subtraction.

Vocabulary

- Sum
- Difference
- Estimate

Material

- Place-value discs
- Appendix p. a20
- Mental Math 10-12

Notes

In *Primary Mathematics* 3 students learned to add and subtract numbers of up to four digits using the standard algorithm and to use rounding to find estimated answers to problems. They also learned how to use bar models to solve word problems. This is reviewed in this chapter. If your student does not know how to fluently and reliably add and subtract 4-digit numbers, you should go back to *Primary Mathematics* 3A to teach this, using place-value discs so he understands the concepts involved.

Two basic bar models are used for solving addition and subtraction word problems, the part-whole model and the comparison model. The part-whole model was discussed in the notes for the previous unit. The comparison model is often used when we are told how many more or less one quantity is than another.

Part-whole model for addition and subtraction:	Comparison model for addition and subtraction:
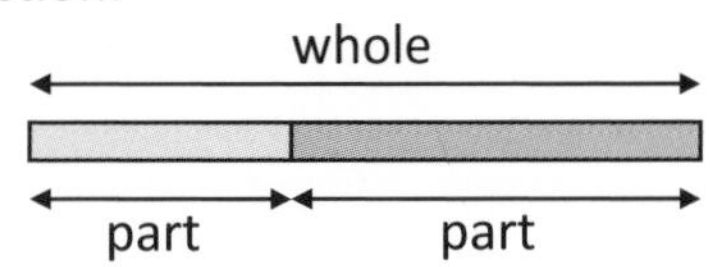	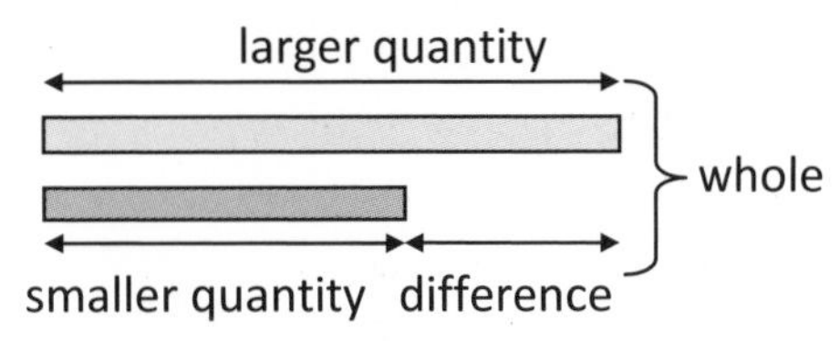
⇒ part + part = whole ⇒ whole – part = part	⇒ larger quantity – smaller quantity = difference ⇒ smaller quantity + larger quantity = whole ⇒ larger quantity – difference = smaller quantity ⇒ smaller quantity + difference = larger quantity

Drawing models helps the student determine what equations are needed to solve the problem. If your student has used earlier levels of *Primary Mathematics*, she will probably be able to solve simple addition or subtraction problems without having to draw a model. Do not insist that she draw models in this chapter if she can solve the problems without it, unless she is new to model drawing. The models will be more useful in problems that also involve multiplication or division. They will be used extensively in *Primary Mathematics* 4 with word problems involving fractions. She needs to be familiar enough with them to draw them when needed, but allowed to skip the pictorial step when she does not need it. Some students can visualize the model in their heads when they are simple, and not need to always draw them out.

It is a good habit for your student to answer word problems in complete sentences so he knows he is answering the question posed by the problem. It is up to you whether you want to require him to answer the problem with a simple sentence or not.

In *Primary Mathematics* 2B, students learned to use mental math strategies to find the difference between a number and 100. Since 100 is the same as 9 tens and 10 ones, we can subtract a number from 100 by mentally thinking of the difference between the tens digit and 9 and then the ones digit and 10.

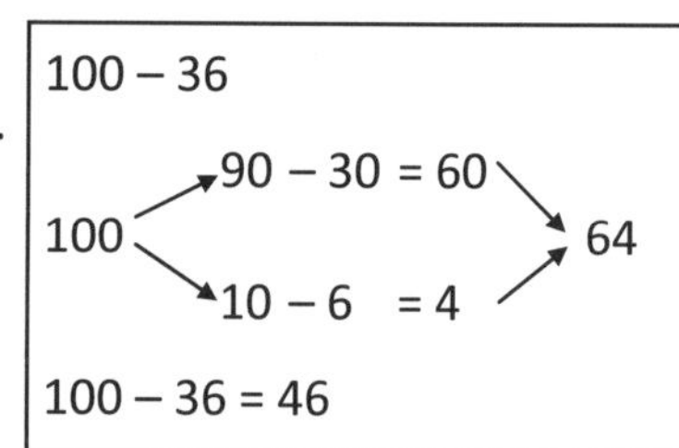

In *Primary Mathematics* 3B, students learned to extend this strategy to subtracting a number from 1000. 1000 is 900, 90, and 10, so we can mentally subtract the hundreds digit from 9, the tens digit from 9, and the ones digit from 10.

1000 – 436

1000 → 900 – 400 = 500; 90 – 30 = 60; 10 – 6 = 4 → 564

1000 – 436 = 564

This strategy is reviewed in this chapter.

Some students will be able to extend these strategies to subtracting a number from a multiple of 100 or 1000.

800 – 642
Subtract 600, and then another 100: 100
Find the difference between 100 and 42: 58
800 – 642 = 158

In *Primary Mathematics* 2, students learned to add a number near 100 by adding 100 and subtracting the difference, and to subtract a number near 100 by subtracting 100 and adding the difference. This strategy is extended to adding or subtracting a number near 1000.

$$\begin{aligned} 149 + 98 &= 149 + 100 - 2 \\ &= 249 - 2 \\ &= 247 \\ 538 - 98 &= 538 - 100 + 2 \\ &= 438 + 2 \\ &= 440 \end{aligned}$$

$$\begin{aligned} 1649 + 998 &= 1649 + 1000 - 2 \\ &= 2649 - 2 \\ &= 2647 \\ 4538 - 998 &= 4538 - 1000 + 2 \\ &= 3538 + 2 \\ &= 3540 \end{aligned}$$

In *Primary Mathematics* 2 and again in *Primary Mathematics* 3, students learned to add and subtract tens and hundreds; that is, numbers with only one non-zero digit, using the strategies learned for adding and subtracting ones. For example, if 32 – 7 = 25, then 3200 – 700 = 2500 and 3216 – 700 = 2516.

If **32** – **7** = **25**,
then **32**00 – **7**00 = **25**00
and **32**16 – **7**00 = **25**16.

4537 + 608
= 4537 + 600 + 8
= 5137 + 8 (45 + 6 = 51)
= 5145 (37 + 8 = 45)

In this chapter, your student will learn to apply the same strategies to adding or subtracting a number with two non-zero digits in steps.

This strategy could be extended to numbers with three non-zero digits, and your student can use it when it is easy to do so. However, do not require her to use mental math on all 3-digit or 4-digit numbers. In some cases regrouping can occur over several place values, such as with the problem 4896 + 725. Adding the ones will affect both the tens and the hundreds, and so adding in steps mentally may take just as long as using the standard algorithm, and the intermediate results are harder to keep track of.

Teach the mental math strategies so that your student understands the concepts, and provide practice, but allow your student the discretion to use the standard algorithm whenever he is uncertain with the mental math and feels more confident of the final answer when using the standard algorithm. The advantage to the standard algorithm is that since operations are done with the lowest place value first (right to left), once the digit in the answer is found, it does not change.

In this chapter, your student will round the numbers in an expression to get an estimated answer. There are times when estimation is the goal, like wanting to know the approximate amount of money needed. In this curriculum, students will mostly use rounding to estimate an answer in order to determine if their calculated answer is reasonable, particularly with regard to place value. This will be particularly useful to check answers when multiplying or dividing by more than 1 digit and in operations with decimals.

There is no particular rule for what place value to round each digit to. It depends on how quickly and accurately the student can do the mental computation with the rounded numbers. For example, for 5707 + 628, we can round both numbers to 1 digit, or both to hundreds. 6000 + 600 = 6600, or 5700 + 600 = 6300. The latter is closer to the exact answer, and the former is easier to solve mentally. In general, your student should round all numbers to the highest place value of the smallest number. In the above example, she should round both numbers to hundreds. But it is not incorrect to round both numbers to one non-zero digit.

(1) Review: Find the sum or difference

Discussion

Concept pages 51-52

Relate the model to the word problem. If needed, show your student how to draw the model rather than simply having him look at the completed model. In the word problem, we are told how many more boys than girls there are. So we use two bars, one longer than the other, for each quantity. We label them with the information in the problem and use a question mark to indicate what we want to find. From the model, we see that we need to subtract to find the number of girls.

You may want to have your student actually work out the two problems on this page to check her understanding of the addition and subtraction algorithms.

Note the use of the words **sum** and **difference** on p. 52. The sum is the answer when two numbers are added together, and the difference is the answer when a smaller number is subtracted from a larger number. The difference is how much more the larger number is than the smaller (or how much less the smaller number is than the larger).

Practice

Tasks 1-2, p. 52

Point out that a part-whole model is used to diagram finding the sum, and a comparison model to diagram finding the difference. When we find the difference between two numbers, we are comparing them to see how much more or less one number is than the other. Have your student find the answers.

1. 1523
2. 207

Provide additional practice as needed. Write down a few random addition and subtraction problems. Include subtraction problems that involve renaming over several place values, such as the one at the right. You can also have your student form random numbers by using 4 sets of number cards 0-10. Shuffle, draw four cards, lay them down in order to make the first number, draw another 3 or 4 to make the second number. An initial 0 means a 3-digit number. Have him find the sum of the two numbers and the difference of the two numbers.

```
 4 10 9
 5  1 0 ¹2
 -    9  4  6
 4  1  5  6
```

Workbook

Exercise 1, pp. 40-41 (answers p. 60)

Enrichment

Tell your student that the same steps used to add and subtract 3-digit and 4-digit numbers can be extended to any number of digits. Have her add and subtract some larger numbers. You can make random numbers using the number cards, or randomly write the digits down for the two numbers.

```
  1   1   1 1
  4 9 0 4 5 1 8
+   8 6 7 0 8 5
  5 7 7 1 6 0 3

    8 9   4
  4 9 0¹4 5¹1 8
-   8 6 7 0 8 5
  4 0 3 7 4 3 3
```

(2) Review: Find the missing number, make 1000

Discussion

Tasks 3-4, p. 53

3. 1201
4. 538

3: Have your student read the problem and relate it to the equation the girl is thinking. Some number – 376 = 825. Ask your student whether we are missing a whole or a part. It has to be a whole since we are subtracting a part to get another part. So in the bar model, the unknown number is the entire bar. The model helps us determine what we use to find the missing number. We add the two parts.

4: Have your student read the problem and determine whether we are being asked to find a missing part or a missing whole. We are missing a part, as shown in the model. So we subtract. Have your student subtract using the standard algorithm. To subtract the ones first, he needs to rename the thousand as 10 hundreds, then one of the hundreds as 10 tens, and then one of the tens as 10 ones. This is the same as renaming 1 thousand as 9 hundreds, 9 tens, and 10 ones. Have your student look at the bottom of p. 53. We can mentally subtract from 1000 by thinking of what we need to add to 4 hundreds to get 9 hundreds, 6 tens to get 9 tens, and 2 ones to get 10 ones.

$$\begin{array}{r} {\scriptstyle 9\ 9\ 10} \\ \not{1}\,\not{0}\,\not{0}\,0 \\ -\ \ 4\ 6\ 2 \\ \hline 5\ 3\ 8 \end{array}$$

Practice

Task 5, p. 54

5. (a) 3	(b) 93	(c) 993
(d) 74	(e) 974	(f) 740

Have your student use mental math to solve these. Point out that to subtract 7 or 26 from 100, we can rename 100 as 9 tens and 10 ones.

Activity

Write the expression **38 + 5** and get your student to solve it mentally using a strategy discussed in the previous unit. Then write the rest of the expressions shown at the right and have him solve each of them mentally. He should see that he can use the answer to 38 + 5 in each, since none of the other place values are affected.

38 + 5 = 43
438 + 5 = 443
2438 + 5 = 2443
3800 + 500 = 4300
3812 + 500 = 4312
3800 + 512 = 4312
380 + 54 = 434

Similarly, write the expression **63 – 7**, have your student solve it, and then have her solve the rest of the expressions shown at the right, using the answer to 63 – 7.

63 – 7 = 56
463 – 7 = 456
8463 – 7 = 8456
6300 – 700 = 5600
6312 – 700 = 5612

Write the expression **2486 + 58** and have your student solve it mentally. Remind him that since 58 is close to 60, one way he can solve this is by adding 60 and subtracting 2.

2486 + 58
= 2486 + 60 – 2
= 2546 – 2
= 2544

Write the expression **100 – 43** and have your student solve it. Then have her use the answer to solve the rest of the problems shown at the right, paying attention to the place-values of the digits.

100 – 43 = 57
/\
90 10

400 – 43 = 357
/\
300 100

3400 – 43 = 3357
1000 – 430 = 570
4000 – 430 = 3570

Write the expression **3850 – 47** and ask your student for ideas on how to solve it mentally. 47 is just 3 less than 50. We can simply subtract 47 from 50. We could also subtract 50 and add 3.

3850 – 47 = 3803
(50 – 47 = 3)

Workbook

Exercise 2, pp. 42-43 (answers p. 60)

Reinforcement

Mental Math 10

(3) Add and subtract mentally

Activity

Write the following expressions and discuss mental math strategies for solving them. The expressions listed here are also on appendix p. a20.

⇒ 7 + 10

$7 + 10 = 17$

⇒ 7 + 9: One method, which we have already learned, is to make a ten with the 9 by taking 1 from the 7. We then have 10 and 6. Point out that 9 is one less than the 10 in the previous expression. Likewise, the answer is one less than 17. Since 9 is the same as 10 – 1, we could add 9 by adding 10 and subtracting 1.

$7 + 9 = 16$

$$\begin{aligned} 7 + 9 &= 7 + 10 - 1 \\ &= 17 - 1 \\ &= 16 \end{aligned}$$

⇒ 37 + 9, 637 + 9, 3637 + 9: Similarly, we can solve these by adding 10 and subtracting 1.

$37 + 9 = 37 + 10 - 1 = 46$

$637 + 9 = 637 + 10 - 1 = 646$

$3637 + 9 = 3637 + 10 - 1 = 3646$

⇒ 637 + 100, 637 + 99, 3637 + 99: Since 99 is one less than 100, we can add 99 by first adding 100 and then subtracting 1.

$637 + 100 = 737$

$637 + 99 = 637 + 100 - 1 = 736$

$3637 + 99 = 3637 + 100 - 1 = 3736$

⇒ 3637 + 1000, 3637 + 999: Since 999 is one less than 1000, we can add 999 by first adding 1000 and then subtracting 1.

$3637 + 1000 = 4637$

$$\begin{aligned} 3637 + 999 &= 4637 + 1000 - 1 \\ &= 4637 - 1 \\ &= 4636 \end{aligned}$$

⇒ 1358 + 7: Show two methods: making a ten, and first adding 10 and then subtracting 3. For the latter, get your student to first tell you how "close" 7 is to 10 (3).

$1358 + 7 = 1365$
(7 split into 2 and 5)

$1358 + 7 = 1358 + 10 - 3 = 1365$

⇒ 1358 + 97, 1358 + 997: Similarly, to add 97, we can add 100 and subtract 3, and to add 997 we can add 1000 and subtract 3.

$1358 + 97 = 1358 + 100 - 3 = 1455$

$1358 + 997 = 1358 + 1000 - 3 = 2355$

⇒ 42 – 10, 42 – 9, 342 – 9, 7342 – 9: We can subtract 9 by subtracting 10 and adding 1. Discuss this strategy. 42 – 9 is one more than 42 – 10. If we subtract 10 instead of 9, we have subtracted 1 too many. So we then have to add back in the 1.

$42 - 10 = 32$

$42 - 9 = 33$

$$\begin{aligned} 42 - 9 &= 42 - 10 + 1 \\ &= 32 + 1 \\ &= 33 \end{aligned}$$

$342 - 9 = 342 - 10 + 1 = 333$

$7342 - 9 = 7333$

⇒ 342 – 100, 342 – 99, 7342 – 99: To subtract 99, we can first subtract 100, then add back in 1.

$342 - 100 = 242$

$342 - 99 = 342 - 100 + 1 = 243$

$7342 - 99 = 7342 - 100 + 1 = 7243$

⇒ 7342 – 1000, 7342 – 999: To subtract 999, we can first subtract 1000, then add back in 1.

7342 – 1000 = 6342
7342 – 999 = 7342 – 1000 + 1 = 6343

⇒ 2345 – 7, 2345 – 97, 2345 – 997: We can subtract the nearest ten (100, 1000) and add back in the difference.

2345 – 7 = 2345 – 10 + 3 = 2338
2345 – 97 = 2345 – 100 + 3 = 2248
2345 – 997 = 2345 – 1000 + 3 = 1348

Remind your student that when he added or subtracted 2-digit numbers, he could do so by first adding or subtracting the tens, and then the ones. You can write some examples, such as **56 + 38** and **62 – 28**. We are adding or subtracting the digits from left to right.

56 + 38 = 86 + 8 = 94
62 – 28 = 42 – 8 = 34

We can also do this with larger numbers. Write the examples **3495 + 1080** and **4509 – 4060** and have your student solve them by first adding or subtracting the thousands, then the hundreds, then the tens, and then the ones. Tell her that it is fairly easy to do this if some of the digits being added or subtracted are small or 0, and there is not going to be too much renaming. In other cases, such as **5687 + 4598** or **6102 – 2398,** it can be easier just to rewrite the problem vertically and use the standard algorithm.

3495 + 1080 = 3495 + 1000 + 80
= 4495 + 80
= 4575

4509 – 4060 = 4509 – 4000 – 60
= 509 – 60
= 449

5687 + 4598

```
   1 1 1
   5 6 8 7
 + 4 5 9 8
 ---------
 1 0 2 8 5
```

6102 – 2398

```
   5 10 9
   6̸ 1̸ 0̸ ¹2
 – 2  3  9  8
 ------------
   3  7  0  4
```

Tell your student that mental math is meant to speed calculations. If he needs to take too long at determining what approach to take, or is not sure of the answer, it is better to use the standard algorithm, particularly if that is less likely to result in an incorrect answer. He can choose whether to use mental math or the standard algorithm on any computation.

Practice

Tasks 6-10, pp. 54-55

6. 1572
7. 2223
8. (a) 2445 (b) 2535 (c) 3435
   (d) 2427 (e) 2337 (f) 1437
9. 3154; 3162
10. 2941; 2939

Workbook

Exercise 3, pp. 44-46 (answers p. 60)

Reinforcement

Mental Math 11-12

Enrichment

You can extend the strategies learned in this lesson to adding and subtracting numbers close to a multiple of 100 or 1000, such as those shown at the right. The process is the same, we just add or subtract the closest 10, 100, or 1000, and then subtract or add the difference.

456 + 48 = 456 + 50 – 2 = 504

502 – 27 = 502 – 30 + 3 = 475

345 + 499 = 345 + 500 – 1 = 844

762 – 398 = 762 – 400 + 2 = 364

4506 + 3998 = 4506 + 4000 – 2 = 8504

9997 – 1999 = 9997 – 2000 + 1 = 7998

(4) Estimate

Discussion

Tasks 11, 13, pp. 55-56

11, 13: We can find an approximate answer for addition or subtraction by rounding each number and then finding the sum or difference of the rounded numbers. That sum or difference is an estimate of the actual answer. Ask your student to find the exact answers as well.

11. 1200
    1200 (1204)
13. 1000
    1000 (1007)

Tell your student that we can use estimated answers to help check if the exact answer is reasonable. If she is good at mental math, there may not seem much point in this for addition and subtraction. Tell her that finding an estimate will become more useful later, primarily in multiplication and division, but also with decimal numbers, which she will learn later.

Tell your student that because an estimation is an approximation, there is no "correct" answer for an estimated answer. Ask him to estimate the value of **7422 – 687** and then find the exact answer. If we round both to numbers where there is only one non-zero digit, we get a different estimated answer than if we round both to the place value where the smallest number has one non-zero digit, in this case hundreds. However, it might be faster to subtract 70 hundreds – 7 hundreds than 74 hundreds – 7 hundreds. There is no rule about what number to round to in order to get an estimate quickly and accurately. (An estimate is not particularly useful if we make a computational error finding the estimate!)

7422 – 687 = 6735
7000 – 700 = 6300
7400 – 700 = 6700

Tasks 15-16, p. 56

There are times when an estimated answer is sufficient, such as in determining the cost of several items. Ask your student for any other situations where an estimated answer is sufficient.

15. $70
    He **has** enough money.
16. 104
    104

Practice

Tasks 12, 14, pp. 55-56

Accept reasonable estimates. Possible answers for estimates are given here. If your student still needs practice with addition and subtraction, you can ask her to find the exact answers as well.

12. (a) 400 + 300 = **700** 384 + 296 = 680
    (b) 500 + 900 = **1400** 507 + 892 = 1399
    (c) 900 + 700 = **1600** 914 + 707 = 1621
    (d) 700 – 400 = **300** 716 – 382 = 334
    (e) 1000 – 300 = **700** 983 – 296 = 687
    (f) 1400 – 700 = **700** 1408 – 693 = 715

14. (a) 400 + 300 + 100 = **800**
    418 + 293 + 108 = 819
    (b) 800 + 600 + 400 = **1800**
    784 + 617 + 399 = 1800
    (c) 800 + 200 – 600 = **400**
    814 + 208 – 587 = 435
    (d) 1200 – 500 – 600 = **100**
    1205 – 489 – 596 = 120

Workbook

Exercise 4, pp. 47-48 (answers p. 61)

(5) Solve word problems

Discussion

Tasks 17-18, p. 57

17. 140 140
18. 2345 2345

Relate the information in the diagram to that in the word problem. If your student is new to *Primary Mathematics*, show the steps for drawing the model.

17: We are given a whole and 2 out of 3 parts, and need to find the third part. So we draw a part-whole model. We label the model with the information in the problem, and use a question mark to indicate what we want to find.

Note that we can solve the problem in different ways. The text shows that we can subtract the men and then the women. We could also add the men and women, and subtract the total adults from the total spectators. Ask your student how we could write a single expression showing the two steps: 6020 – (3860 + 2020).

You can point out that subtracting two quantities that have been added together is the same as subtracting each one separately. We cannot simply remove the parentheses and subtract the men and then add the women. If we remove the parentheses, we have to show that we need to subtract both quantities.

18: In this problem, we are comparing the number of chickens and ducks, so we can draw a comparison model. For a comparison model, one quantity is larger than the other and we draw that one with a longer bar. Then we label the bars with the information in the problem, and what we want to find with a question mark. If you are showing your student how to draw the model, the question mark would go on the total, since that is what we need to find. From the model, we see that we need to first find the number of chickens.

Ask your student to write a single expression: 1025 + 295 + 1025. Point out that we can divide the chicken bar into two parts where one part is equal to the bar for the ducks. Then the chicken bar is the same as the duck bar plus how many more chickens there are than ducks. We could write the equation 1025 + 1025 + 295 (duck + duck + difference).

Practice

Practice A, p. 58

1-6: Allow your student to decide which of these to do using mental math, and which to do using the standard algorithm.

7: Estimated answers can vary. If your student needs more practice with addition and subtraction, you can ask him to also find the exact answer.

	(a)	(b)	(c)
1.	530	1651	1574
2.	543	557	856
3.	377	108	429
4.	354	536	476
5.	3465	6019	8080
6.	2157	300	3017

7.
	Estimate	Exact
(a)	600 + 300 = 900	(905)
(b)	2200 + 900 = 3100	(3041)
(c)	3900 + 200 = 4100	(4156)
(d)	700 – 200 = 500	(475)
(e)	7100 – 200 = 6900	(6840)
(f)	5400 – 200 = 5200	(5223)
(g)	2600 + 100 – 500 = 2200	(2215)
(h)	1400 – 900 – 300 = 200	(160)

8-12: Allow your student to decide whether to draw a model for these or not. She is not likely to need a model with these problems if she has done earlier levels of *Primary Mathematics*. If she needs more practice with estimation, you can ask her to find an estimate first.

8. **Alan** collected more.
   243 – 174 = **69**
   Alan collected 69 more matchboxes.
9. 438 + 15 = **453**
   Helen scored 453 points.
10. $2785 + $536 = **$3321**
    The motorcycle costs $3321.
11. $99 + $286 = **$385**
    She had $385 at first.
12. 6345 – 3016 – 2107 = **1222**
    There are 1222 green beads.

Workbook

Exercise 5, pp. 49-50 (answers p. 61)

Note: Problem 1 uses the words "more than" to compare two quantities, and problem 2 uses the words "less than" to compare two quantities. Your student may simply look at the words and use subtraction in the second problem. Both, however, are solved with addition. Your student cannot look for key words to solve word problems in this curriculum; he needs to read the problems carefully, and possibly model them.

Reinforcement

Extra Practice, Unit 2, Exercise 1, pp. 23-24

Tests

Tests, Unit 2, 1A and 1B, pp. 33-36

Workbook

Exercise 1, pp. 40-41

1. (a) 7895 (b) 3009 (c) 8954
   (d) 3054 (e) 11,044 (f) 12,054
   (g) 14,054 (h) 14,300 (i) 18,492
   (j) 16,854 (k) 10,015 (l) 9212

2. (a) 4444 (b) 7456 (c) 3889
   (d) 126 (e) 8873 (f) 2373
   (g) 354 (h) 109 (i) 7888
   (j) 2668 (k) 983 (l) 1069

Exercise 2, pp. 42-43

1. (a) 10, 30, 330, 6330
   (b) 10, 60, 260, 3260
   (c) 100, 600, 1100, 2600
   (d) 100, 500, 2100, 5500
   (e) 1000, 3000
   (f) 1000, 6000

2. (a) 2461
   (b) 2845
   (c) 3553

3. (a) 4, 34, 244, 5224
   (b) 1, 91, 991, 1991
   (c) 28, 428, 928, 2028
   (d) 1, 301, 2001, 2901
   (e) 160, 1160
   (f) 1, 2001

4. (a) 2502
   (b) 3401
   (c) 3503

Exercise 3, pp. 44-46

1. (a) 3904
   (b) 3640
   (c) 4302

2. (a) 3024
   (b) 3200
   (c) 3273

3. (a) 1591, 1681, 2581
   (b) 2764, 2854, 3754

4. (a) 2020
   (b) 2602
   (c) 3146
   (d) 3801
   (e) 4102

5. (a) 2600
   (b) 2250
   (c) 2470
   (d) 2001
   (e) 3002

6. (a) 2603; 2611
   2611
   (b) 5276; 5316
   5316
   (c) 2924; 2894
   2894
   (d) 2261; 1761
   1761

7. (a) 4625
   (b) 7426

8. (a) 2816
   (b) 2997

Workbook

Exercise 4, pp. 47-48

1. (a) 900 (908)
   (b) 300 (273)
   (c) 800 + 200 = 1000 (996)
   (d) 900 – 300 = 600 (615)
   (e) 600 + 600 = 1200 (1201)
   (f) 900 – 300 = 600 (564)
   (g) 1800 + 400 = 2200 (2214)
   (h) 2300 – 1000 = 1300 (1308)

2. (a) 800 (799)
   (b) 700 – 200 – 300 = 200 (196)
   (c) 1000 – 200 + 100 = 900 (901)
   (d) 500 + 300 – 300 = 500 (506)
   (e) 2000 – 600 + 500 = 1900 (1874)
   (f) 2400 + 600 – 700 = 2300 (2294)
   (g) 1100 – 100 + 400 = 1400 (1403)
   (h) 3000 + 1000 + 400 = 4400 (4410)

Exercise 5, pp. 49-50

1. 
4670 | 698
European
African
?

4670 + 4670 + 698 = **10,038**
She has 10,038 stamps.

2.
325 | 49
Saturday
Sunday
?

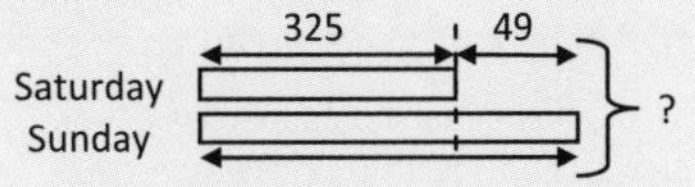

325 mi + 325 mi + 49 mi = **699 mi**
He drove 699 miles on the two days.

3. $3225 – $1950 – $625 = **$650**
She needs $650.

4.
$2365 | $375
January
February

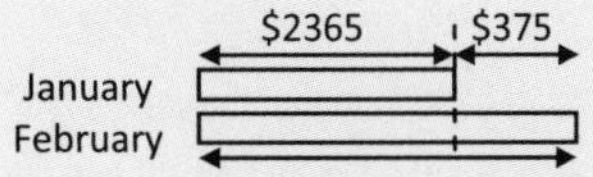

Total earned: $2365 + $2365 + $375 = $5105
Amount saved: $5105 – $4250 = **$855**
He saved $855.

Chapter 2– Multiplication and Division

Objectives

- Review multiplication and division of a number within 10,000 by a 1-digit number.
- Use approximation to estimate the product or the quotient.
- Review modeling method for solving word problems involving multiplication and division.
- Multiply a number within 10,000 by a 2-digit number.

Vocabulary

- Product
- Quotient
- Remainder

Material

- Place-value discs

Notes

In *Primary Mathematics* 3A, students learned to multiply and divide a 3-digit or a 4-digit whole number by a 1-digit whole number. In this chapter, the multiplication and division algorithms are reviewed.

If your student has done earlier levels of *Primary Mathematics*, he should already understand the process and the concepts behind the standard algorithms for multiplication and division of a multi-digit number by a 1-digit number. In the U.S. students often do not go beyond simple division facts and multiplying a 2-digit number by a 1-digit number in the third grade. If your student is new to the concepts being reviewed here, it is highly recommended that you go back to *Primary Mathematics* 3A and teach the algorithm following the clear and straightforward approach used there, thus helping your student master a difficult topic with relative ease. Then he will not need to learn and memorize the steps to various alternate, paper-and-pencil strategies that do not extend well to larger numbers and require considerably more steps to follow, such as a lattice method or an area model or a partial product method, that are prevalent in some U.S. texts.

If your student has used earlier levels, but needs a more concrete review than simply following the steps on paper, use place-value discs to illustrate the algorithm. The steps for an example problem will be shown at the end of these notes.

Students learned the terms **product**, **quotient**, and **remainder** in *Primary Mathematics* 3A. These are reinforced here. You can use the terms **factor** and **multiple** now as well.

In *Primary Mathematics* 3A, students learned to use approximation to make rough estimates of the product or quotient. This will be reviewed here. Students are encouraged to estimate their answers mentally in advance. Being able to make estimates helps them check for errors in place value, and allows them to build a useful "feeling" for numbers. Checking for errors in place value becomes even more necessary when multiplying and dividing by a 2-digit number, and in multiplying and dividing decimals.

To estimate the answer for multiplication of numbers with 2 or more digits, we round to convenient multiples of 10, 100, or 1000.

382 x 8
↓
400 x 8 = 3200

To estimate division of numbers with 3 or more digits, we round the dividend to a multiple of 10, 100, or 1000 so that the first two digits are a close multiple of the divisor. In the example shown here, we can round 4812 to 4900, not 5000,

4812 ÷ 7
↓
4900 ÷ 7 = 700

since 49 is a multiple of 7. Rounding to 5000 would not help in finding an estimate, since it is not divisible by 7.

For more continuous practice, you may want to give your student several problems involving multiplication and division, and sometimes addition and subtraction, periodically throughout the succeeding units.

In *Primary Mathematics* 3, students learned to apply the part-whole and comparison models to situations involving multiplication or division.

For a part-whole model for multiplication and division, we now draw equal parts. These equal parts are called units, and in solving the problems, we generally need to find the value of 1 unit. We can see the relationship among the three quantities: the whole, one part, and the number of parts.

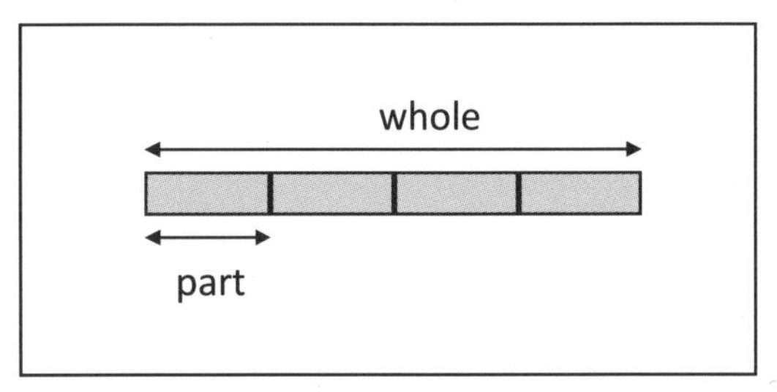

One part x number of parts = whole

Whole ÷ number of parts = one part

Whole ÷ one part = number of parts

For a comparison model for multiplication and division, we also use units. Two quantities are compared such that one quantity is a multiple of the other.

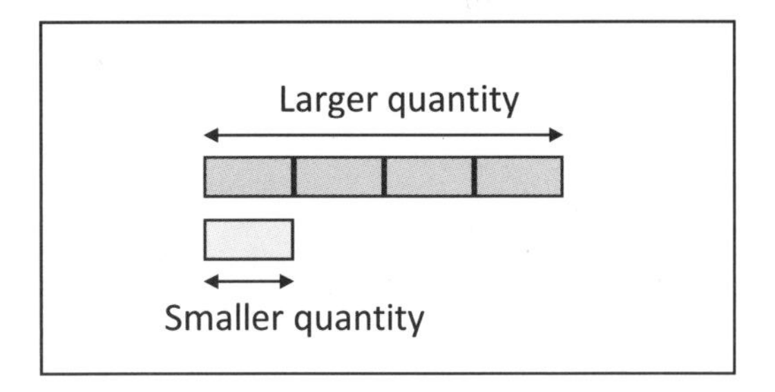

Larger quantity ÷ smaller quantity = multiple

Smaller quantity x multiple = larger quantity

Larger quantity ÷ multiple = smaller quantity

These models are a powerful tool when used with problems that involve more than one step and a combination of operations. For example:

⇒ The total length of 4 strings is 100 cm. String A is 9 cm shorter than string B. String B is three times as long as string C. String D is 13 cm longer than string C. How long is the longest string?

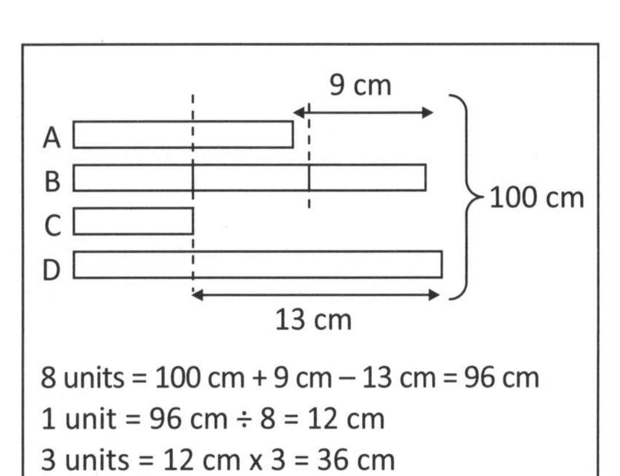

8 units = 100 cm + 9 cm – 13 cm = 96 cm
1 unit = 96 cm ÷ 8 = 12 cm
3 units = 12 cm x 3 = 36 cm
The longest string is 36 cm long.

Since we are told that string B is three times as long as string C, draw strings B and C first. We do not know at first whether string A is more than a unit shorter than string B, or whether string D is more than a unit longer than string C, or even whether it might be the longest string, but how we draw A and D will not change the method used to solve this problem. If we add 9 cm to A, and subtract 13 cm from D, then we will have 8 units. After we find that 1 unit is 12 cm, we know that B is the longest string, not D.

Bar models are a problem-solving tool. They are a means to an end, not the end itself. The goal in drawing models is to solve the problem, not to draw models. Do not require your student to draw models if she can easily solve the problem without them. However, she should be capable of drawing and using models when needed.

Although your student needs to be able to use models effectively, learning to use them should not involve learning a set of imposed steps for drawing them. There is no set of steps that will apply to all types of problems he will encounter, and imposing a set of specific steps short-changes development of problem-solving skills and the logic needed to solve word problems. A set of steps that might be devised for simpler problems will not be easily applied to more challenging problems he will encounter later.

It is not sufficient, for example, to tell your student to draw a unit for each thing the problem is about, and then add units or parts or labels while going sentence by sentence through the problem. If the problem does not fit those steps, either the steps will not work, or a lot of back-tracking will have to be done. In some cases, drawing the model starting with information in some part of the problem other than the first sentence results in a simpler model that is easier to interpret. With experience with a wide variety of problems, rather than a specified set of steps, your student will become adept at manipulating the basic types of models in a way that will best help her to solve the problem, and will be able to, if necessary, attempt several approaches if one approach does not easily lead to a solution.

Using place-value discs to illustrate 4576 x 4:

4576 is made up of 4000, 500, 70, and 6. We will multiply each of these parts by 4, starting with the lowest place value.

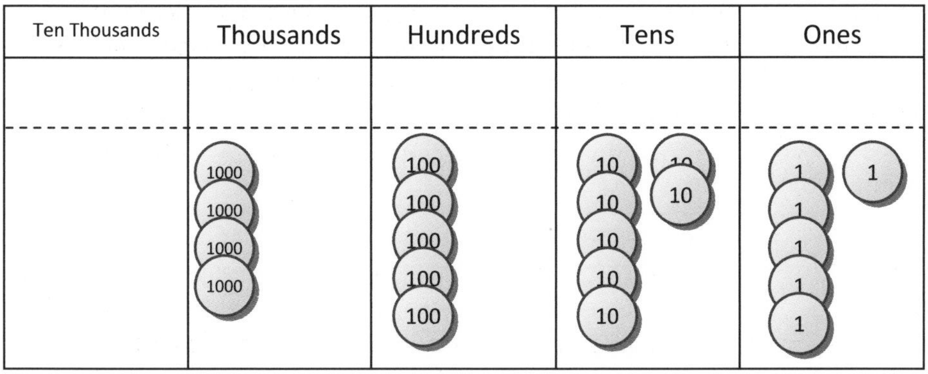

Ten Thousands	Thousands	Hundreds	Tens	Ones
	1000 1000 1000 1000	100 100 100 100 100	10 10 10 10 10 10 10	1 1 1 1 1 1

```
 4 5 7 6
x      4
```

First, we multiply the ones by 4. 6 x 4 = 24 (Move the 6 ones off the chart and quadruple them.) In order to put them on the chart, we rename 20 as 2 tens. (Replace twenty 1-discs with two 10-discs, and put the renamed tens at the top of the tens column, above the dotted line, and the remaining ones in the ones column.) To show this step on the written work, we write a little 2 above the tens to remind us we have 2 tens from multiplying the ones by 4 and write the ones below the line in the ones place.

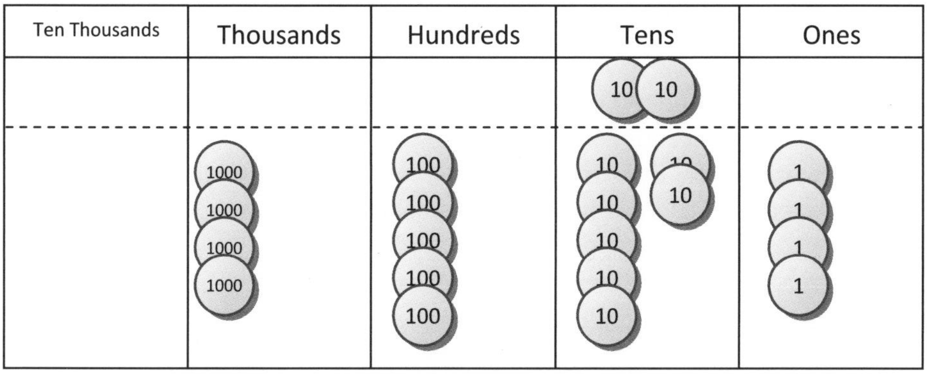

Ten Thousands	Thousands	Hundreds	Tens	Ones
			10 10	
	1000 1000 1000 1000	100 100 100 100 100	10 10 10 10 10 10 10	1 1 1 1

```
4000
 500
  70
   6 x 4     = 24

       2
 4 5 7 6
x      4
       4
```

Next, we multiply the original 7 tens by 4. 70 x 4 = 280 (7 tens x 4 = 28 tens. Take the seven 10-discs off the chart and quadruple them.) We now have 28 tens, plus the 2 tens (from above the dotted line), so we have a total of 30 tens. To place them all on the chart, we need to rename them as 3 hundreds. (Remove the two tens left above the dotted line and put 3 hundreds at the top of the hundreds column, above the dotted line.) To show this step on the written work, we write a little 3 above the hundreds place and a 0 below the line in the tens place.

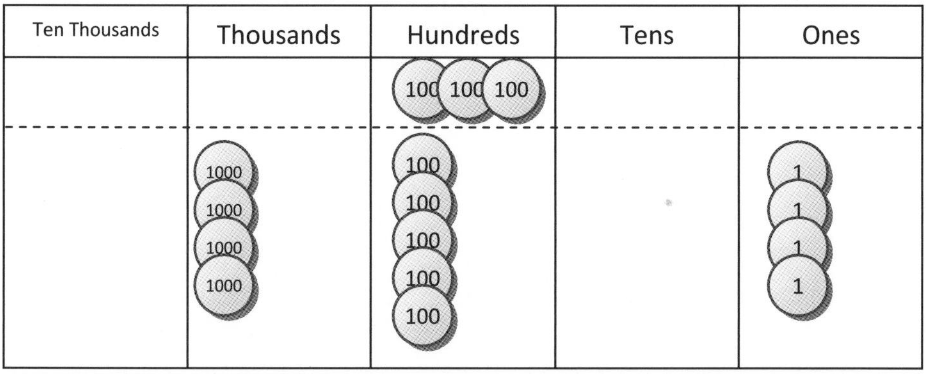

Ten Thousands	Thousands	Hundreds	Tens	Ones
		100 100 100		
	1000 1000 1000 1000	100 100 100 100 100		1 1 1 1

```
4000
 500
  70 x 4 =   280
   6 x 4 =    24

     3 2
 4 5 7 6
x      4
     0 4
```

Next, we multiply the hundreds by 4. Again, we just multiply the original 5 hundreds; we don't include the 3 hundreds (above the dotted line) that came from multiplying the tens by 4 since they are not the hundreds of the original number, 4576. 500 x 4 = 2000 (5 hundreds x 4 = 20 hundreds. Take the five 100-discs off the chart and quadruple them.) We now have 20 hundreds, plus the 3 hundreds from multiplying the tens, so we have a total of 23 hundreds. To put them all on the chart, we rename 20 of them as 2 thousands. (Replace twenty 100-discs with two 1000-discs**,** place these in the thousands column above the dotted line, and place the remaining three 100-discs in the hundreds column.) We show this on the written work by writing a little 2 above the thousands place and a 3 below the line in the hundreds place.

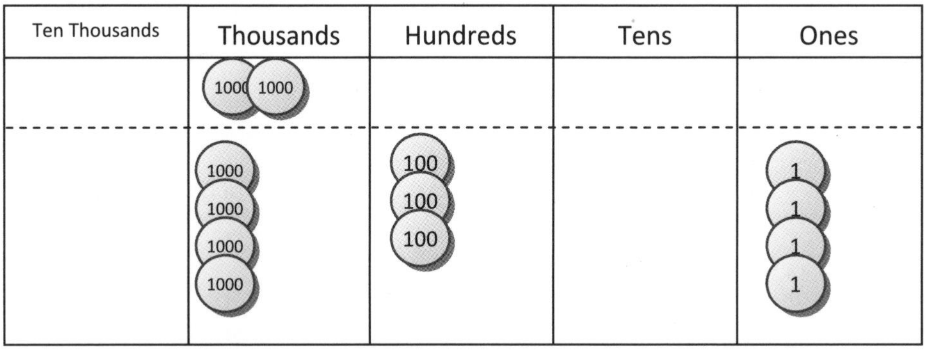

Ten Thousands	Thousands	Hundreds	Tens	Ones
	1000 1000			
	1000 1000 1000 1000	100 100 100		1 1 1 1

4000
500 x 4 = 2000
70 x 4 = 280
6 x 4 = 24

```
  2 3 2
  4 5 7 6
x       4
    3 0 4
```

Finally, we multiply the original thousands by 4. 4000 x 4 = 16,000 (4 thousands x 4 = 16 thousands. Take the four 1000-discs off the chart and quadruple them.) Adding the 2 thousands we have from multiplying the hundreds, we now have 18 thousands. To place them all on the chart, we need to rename them as 1 ten thousand and 8 thousands. (Replace ten 1000-discs with a 10,000 disc. Place the 10,000-disc in the ten thousands column, and the eight 1000-discs in the thousands column.) We show this on the written work by writing 18 in the ten thousands and thousands places. Since we have no more place values to multiply, we can include the 1 ten thousands under the line in the answer.

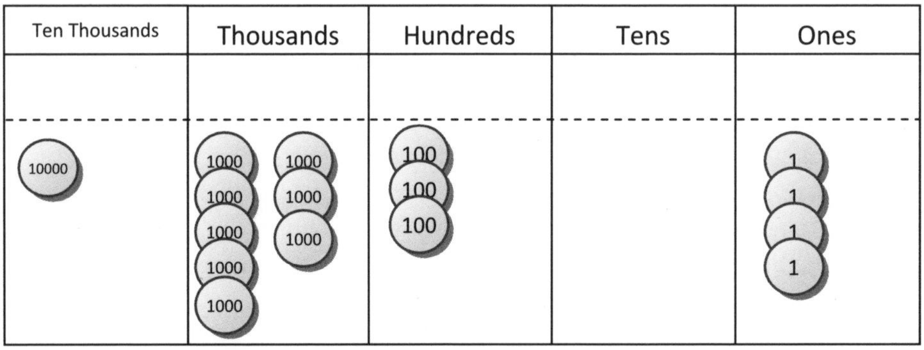

Ten Thousands	Thousands	Hundreds	Tens	Ones
10000	1000 1000 1000 1000 1000 1000 1000 1000	100 100 100		1 1 1 1

4000 **x 4 = 16000**
500 x 4 = 2000
70 x 4 = 280
6 x 4 = 24

```
    2 3 2
    4 5 7 6
x         4
  1 8 3 0 4
```

We have essentially multiplied each place value by 4, and added all the products together as we went along.

Note that multiplying the tens did not change the final ones, and multiplying the hundreds did not change the final tens. The advantage of the standard algorithm for multiplication, i.e., working from the lowest place value (right to left) is that once the answer is written below the line, it does not change. It is possible to multiply starting with the greatest place value (left to right), but we need to record or remember some digits that might change after multiplying the lower place values.

Using place-value discs to illustrate 5508 ÷ 3:

Division can involve either the process of sharing (e.g. put 5508 into 3 equal parts and find the number in each part) or grouping (e.g. group 5508 by 3's to find the number of groups; that is, the number of 3's in 5508). To illustrate division with place-value discs, we use a sharing process. In *Primary Mathematics* 3A, students learned that since grouping gives the same answer as sharing, we can use the same process as if we were sharing instead. It would be time consuming to make groups of three.

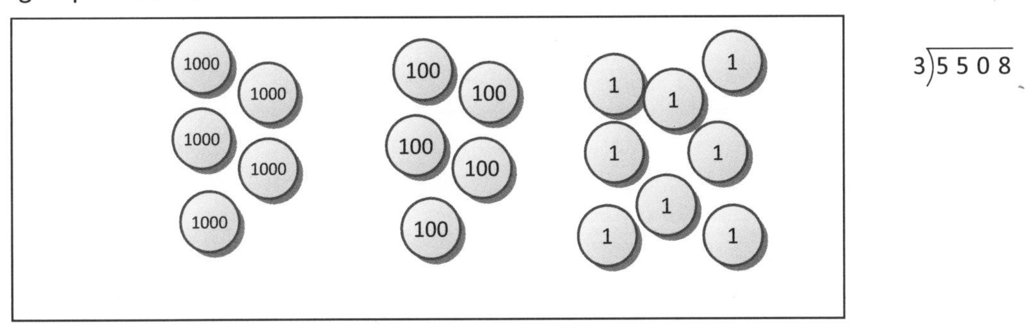

With the division symbol, $\overline{)\quad}$ the number being divided (the dividend) goes under the line, and the number we are dividing by (the divisor) goes to the left of the curved part. The answer (the quotient) goes above the line. We show our work under the number being divided.

For division, we always start with the highest place value and continue from there. Here, we first put the thousands into equal groups. (Put the 1000-discs into 3 groups, with 2 left over.) We show this on the written format by writing the amount in each group (the quotient) above the line. Below the dividend, we write how many were used (1000 x 3 = 3000, 3 thousands were used), and subtract to find the remainder. The remainder should not be greater than the divisor (3). We align the written work both above and below the line. That way we don't have to add extra 0's to show what place value we are working on at each step. We cannot divide the remaining 2 thousands equally into the three parts, so we rename them as 20 hundreds. (Replace the two 1000-discs with twenty 100-discs.) We now have a total of 25 hundreds. We show this on the written work by "bringing down" the 5 in the hundreds place.

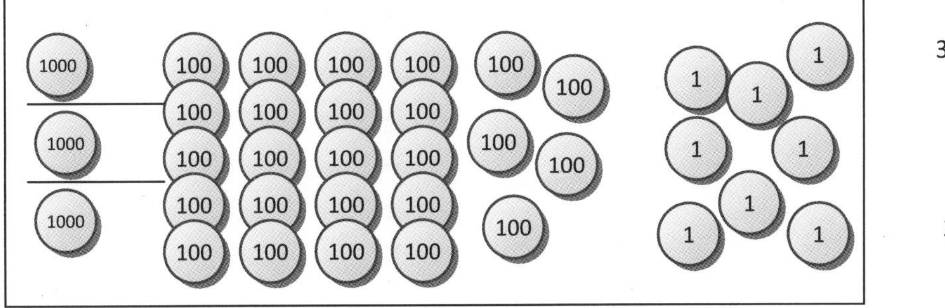


```
   1
3)5 5 0 8
  3        ← 1000 x 3
  2        ← 5000 – 3000

      ↓

   1
3)5 5 0 8
  3
  2 5
```

Next, we divide the 25 hundreds into the three parts. 25 hundreds ÷ 3 = 8 hundreds with 1 hundred left over. We show this on the written work by writing 8 (the quotient, how many are in each group) above the line, the amount used, 24 hundreds (800 x 3 = 2400) below, and then subtract to find the remainder. Again, we cannot divide the remaining hundred into the three parts, so we rename it as 10 tens. (Replace the remaining 100-disc with ten 10-discs.) We now have a total of 10 tens. We show this on the written work by "bringing down" the 0 in the tens place.

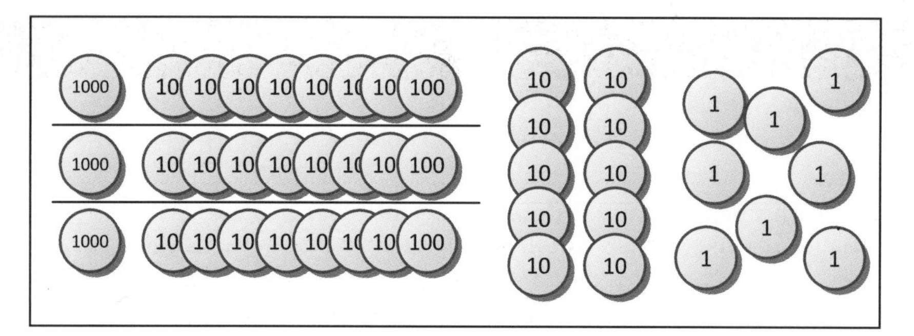

```
   1 8
3)5 5 0 8
  3
  2 5
  2 4
    1 0
```

Next, we divide the 10 tens into the three parts. 10 tens ÷ 3 = 3 tens with 1 ten left over. We show this on the written work by writing 3 (the quotient) above the line, the amount used, 9 tens, below, and then subtracting to find the remainder of 1. Again, we cannot divide the remaining ten into the three parts, so we rename it as 10 ones. We now have a total of 18 ones. We show this on the written work by "bringing down" the 8 in the ones place.

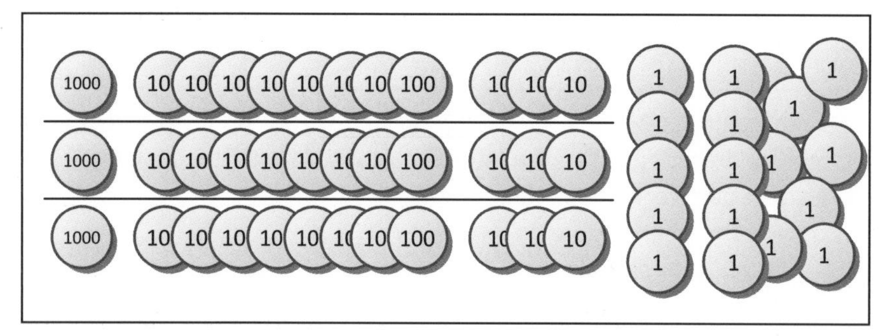

```
   1 8 3
3)5 5 0 8
  3
  2 5
  2 4
    1 0
      9
      1 8
```

Finally, we divide the 18 ones into the three parts. We show this on the written work by writing 6 (the quotient) above the line, the amount used, 18 ones, below. There are no remainders.

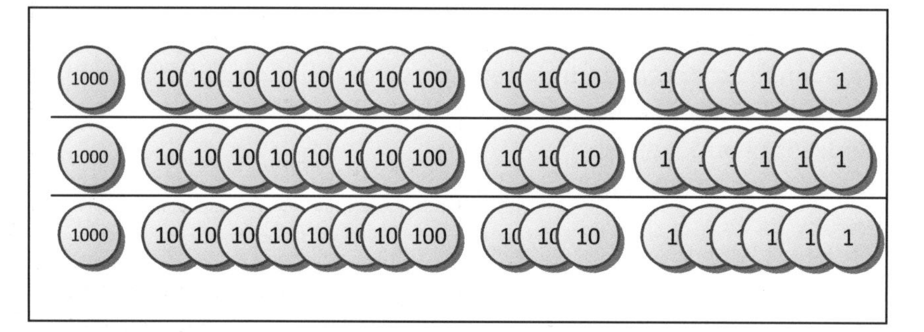

```
   1 8 3 6
3)5 5 0 8
  3
  2 5
  2 4
    1 0
      9
      1 8
      1 8
        0
```

We are essentially splitting the number up into thousands, hundreds, tens, and ones that are divisible by 3.

```
   1 8 3 6        5508   ÷ 3
3)5 5 0 8
  3          ←    3000   ÷ 3 =   1000
  2 5
  2 4        ←    2400   ÷ 3 =    800
    1 0
      9      ←      90   ÷ 3 =     30
      1 8
      1 8    ←      18   ÷ 3 =      6
        0         5508          1836
```

(1) Review: Multiplication

Discussion

Concept p. 59

Have your student read the word problem out loud. Relate the bar model to the information in the word problem. Since we are told that there are three times as many foreign stamps as U.S. stamps we use a comparison model. We show the amounts with equal units to show three times as many foreign stamps. Since all the parts are equal, we know that the problem will involve multiplication or division. Since we want to find the total, given the number in one part, we use multiplication.

Have your student work through the multiplication problem on paper, explaining each step. If he cannot do the problem, then you may need to review the algorithm. You may want to do some easier problems, involving multiplication of first 2-digit numbers and then 3-digit numbers, showing the steps with place-value discs. This can take an extra lesson.

Tasks 1-3, p. 61

> 2. (a) 17,700 (b) 25,960
> 3. 24,000
> 24,000

3: Tell your student that, as with addition and subtraction, we can estimate the answer to multiplication problems by rounding the number to the closest multiple of 10 (100, 1000) such that the rounded number has only one non-zero digit. Then we can find the estimated product mentally.

Use the example at the right for the following. In any estimation, we want to get an estimated answer as close to the exact answer that we can calculate quickly. Sometimes, if the mental computation is easy, we can round to two non-zero digits. We will get a closer estimate, particularly if the number is in the middle, between tens or hundreds or thousands.

> 1549 x 3
> ↓
> 1500 x 3 = 4500
> Or
> 1549 x 3
> ↓
> 2000 x 3 = 6000
>
> 1549 x 3 = 4647

Practice

Task 4, p. 61

> 4. (a) 4000 x 5 = 20,000 20,380
> (b) 4000 x 8 = 32,000 34,536
> (c) 2000 x 9 = 18,000 18,450
> (d) 7 x 7000 = 49,000 48,517
> (e) 9 x 2000 = 18,000 19,557
> (f) 6 x 4000 = 24,000 23,040

Workbook

Exercise 6, p. 51 (answers p. 81)

Enrichment

Have your student do the following problems and discuss any patterns she might observe.

9999 x 1	9999 x 6
9999 x 2	9999 x 7
9999 x 3	9999 x 8
9999 x 4	9999 x 9
9999 x 5	9999 x 10

(2) Review: Division

Discussion

Concept p. 60

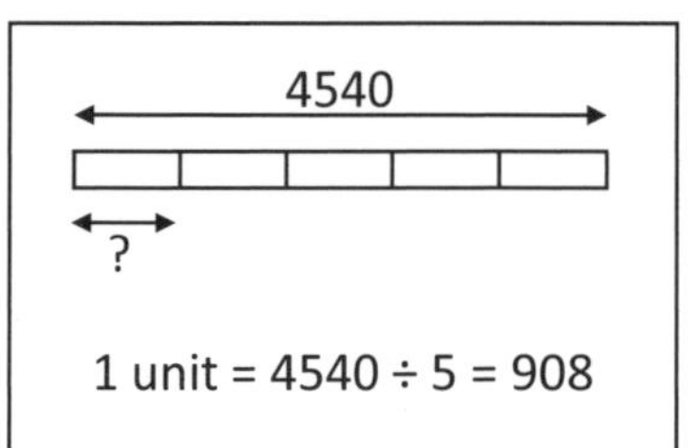

Have your student read the word problem out loud. Remind her that we can also draw models for division problems. Show a part-whole model. Each packet of stamps is the same size, so we draw equal units. This time, we are given the whole and want to find the value of one unit. (This particular problem is easy enough to understand so a model probably isn't needed to determine how to solve it.)

Have your student work through the division problem on paper, explaining each step. If she cannot do the problem, then you may need to review the algorithm. You may want to do some easier problems, involving division of first 2-digit numbers (e.g. 53 ÷ 2) and then 3-digit numbers (e.g. 867 ÷ 5), showing the steps with place-value discs. This can take an extra lesson.

Tasks 5-6, p. 62

6. (a) 650 (b) 605

Activity

In Task 7, your student will divide by 10. You may want to give him a concrete introduction first.

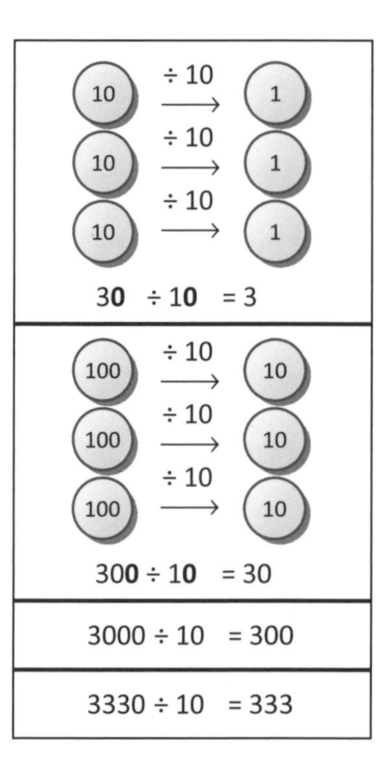

Give your student 3 tens and ask her to divide them into 10 equal groups. She may try to trade in each ten for ones until she has 30 ones and then divide those up into 10 piles of 3 each. So 30 ÷ 10 = 3. Tell her she can do the same thing in steps by dividing each ten individually by 10. Each ten would put 1 into each of the ten groups. Since there are 3 tens, each group would end up with 3 ones. There are the same number of ones in each group as tens.

Now give your student 3 hundreds and ask him how he could divide 300 by 10. Each hundred is 10 tens, so each would contribute one ten to each of the 10 groups. Each group ends up with 3 tens, or 30. So 300 ÷ 10 = 30.

Repeat with 3 thousands and ask your student to divide by ten.

Now have your student start with 3 thousands, 3 hundreds, and 3 tens and divide by 10. As before, each 1000-disc, when divided by 10, yields a 100-disc to each of the ten groups, each 100-disc yields a 10-disc, and each 10-disc yields a 1-disc, so that ten each group of the 10 groups now has 3 hundreds, 3 tens, and 3 ones.

If a number ends in a 0, we can divide it by 10 by simply removing the 0, "sliding" the place value over one place.

Write down and have your student do the following problems:

⇒ 60 ÷ 10	(6)
⇒ 400 ÷ 10	(40)
⇒ 320 ÷ 10	(32)
⇒ 8930 ÷ 10	(893)
⇒ 4300 ÷ 10	(430)
⇒ 6000 ÷ 10	(600)
⇒ 23,980 ÷ 10	(2398)
⇒ 29,100 ÷ 10	(2910)
⇒ 15,000 ÷ 10	(1500)
⇒ 30,000 ÷ 10	(3000)
⇒ 100,000 ÷ 10	(10,000)

Discussion

Tasks 7-9, pp. 62-63

7. (a) 4
   (b) 40
   (c) 44
   (d) 444

8. 600
   600

8: Ask your student why we round 3840 to 3600 rather than 4000. We want to find the estimate easily, and dividing 4000 by 6 is not any easier than dividing 3840 by 6. We round to the nearest number that we can easily divide evenly.

Ask your student to find an estimate for 3758 ÷ 8. This time we would round to the nearest 4000, but only because 40 hundreds can be divided easily by 8.

3758 ÷ 8
↓
4000 ÷ 8 = 500

3758 ÷ 8 = 469 R 5

9: After going through this task, ask your student if she notices anything about the quotient. It is the same as the first three digits of 3245. We can split 3245 into 3240 and 5. 3240 is easily divided by 10 by removing the 0, and 5 cannot be divided by 10, and so it will be the remainder.

3245
/\
3240 5

3240 ÷ 10 = 324
3245 ÷ 10 = 324 R 5

Practice

Task 10, p. 64

10.	(a) 3600 ÷ 9 = 400	400 r4
	(b) 3500 ÷ 7 = 500	511 r3
	(c) 3200 ÷ 8 = 400	390
	(d) 8000 ÷ 10 = 800	812 r8
	(e) 6000 ÷ 3 = 2000	2509 r1
	(f) 6000 ÷ 6 = 1000	1196 r4

Workbook

Exercise 7, pp. 52-53 (answers p. 81)

Enrichment

Have your student do the following problems and discuss any patterns he might observe.

1000 ÷ 9	5000 ÷ 9
2000 ÷ 9	6000 ÷ 9
3000 ÷ 9	7000 ÷ 9
4000 ÷ 9	8000 ÷ 9

(3) Review: Word Problems

Discussion

Tasks 11-14, pp. 64-66

Tasks 12 and 13 are not modeled. Not all problems need to be modeled, and there isn't much advantage to modeling these two unless your student has not done earlier levels of *Primary Mathematics* and needs more practice with modeling the problems. Task 14 is an example of a situation where an estimate is sufficient.

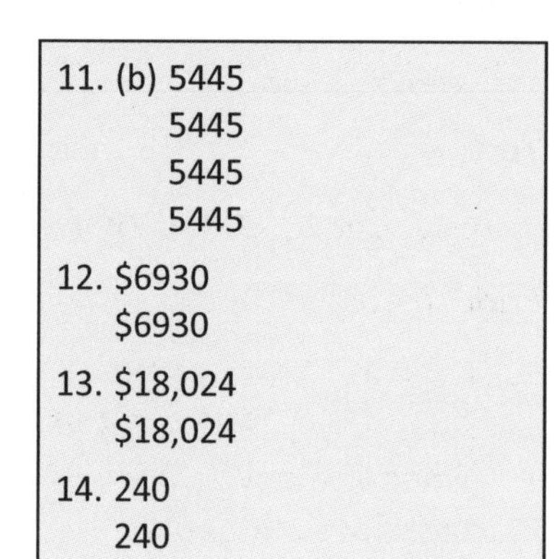
11. (b) 5445
5445
5445
5445
12. $6930
$6930
13. $18,024
$18,024
14. 240
240

Activity

Discuss some additional word problems where it is helpful to model the problem, such as the following. Guide your student in modeling these.

⇒ John and Matt earn the same amount of money. If John spends $130 and Matt spends $480, John will have three times as much money left as Matt. How much money does each boy earn?

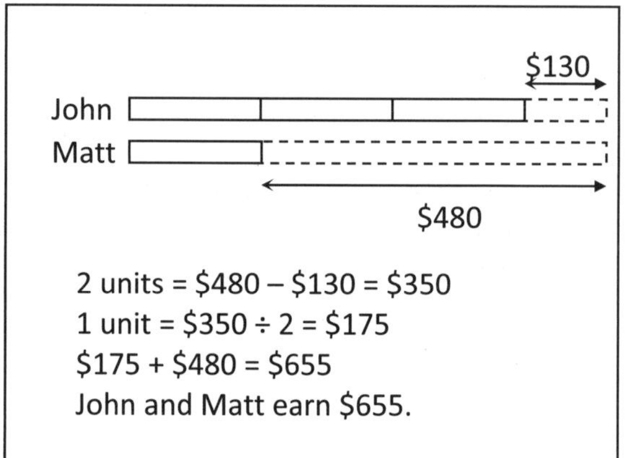

2 units = $480 – $130 = $350
1 unit = $350 ÷ 2 = $175
$175 + $480 = $655
John and Matt earn $655.

⇒ Amy has 34 stickers and Sara has 20. Amy gives Sara some stickers so they both have the same number of stickers. How many stickers does Amy have now? How many stickers did Amy give to Sara?

The total number of stickers does not change. Amy ends up with 1 unit, and Sara ends up with 1 unit.

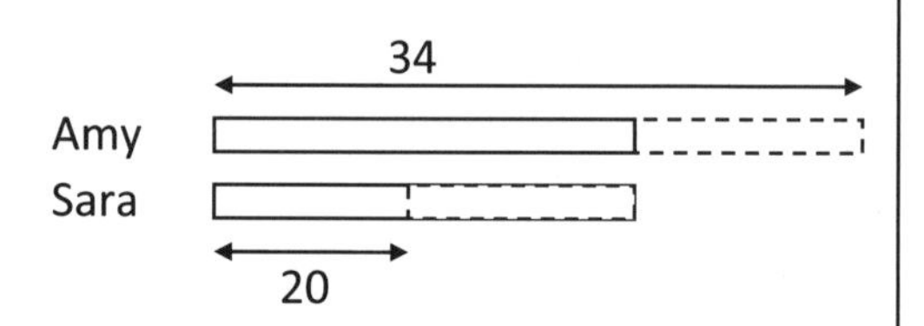

2 units = 34 + 20 = 54
1 unit = 54 ÷ 2 = 27
34 – 27 = 7
Or:
Amy has 34 – 20 = 14 stickers more than Sara. If she gives half of them, or 7, to Sara, they will have the same number, 27 stickers.
Amy gives Sarah 7 stickers.

Workbook

Exercise 8, pp. 54-55 (answers p. 81)

Reinforcement

Extra Practice, Unit 2, Exercise 2, pp. 25-26

(4) Practice

Practice

Practice B, p. 67

Tests

Tests, Unit 2, 2A and 2B, pp. 37-42

Enrichment

⇒ Amy and Zoe have the same number of coins. After Amy gave Zoe 24 of her coins, Zoe had 4 times as many as Amy. How many coins do they have altogether?

Because we are told how many more one has than the other after the transfer of coins, we can draw that first.

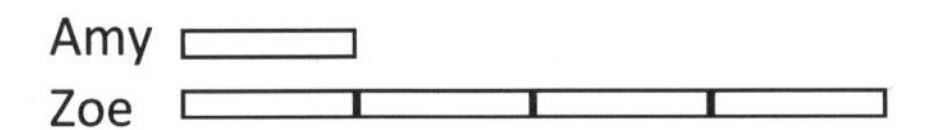

Now we need to move some of the bar from Zoe to Amy to get the equal amount they started with. To do that, we need half the difference. That will be one and a half units. To avoid half-units, we can divide all the units in half.

So now we can move three units to Amy, showing that they both started with the same number of coins.

Amy
Zoe

Amy gave 3 units to Zoe, so 3 units must be 24. Altogether, they have 10 units.

3 units = 24
1 unit = 24 ÷ 3 = 8
10 units = 8 x 10 = 80

They have 80 coins altogether.

1. (a) 6033	(b) 8428	(c) 17,250
2. (a) 25,290	(b) 27,419	(c) 56,322
3. (a) 703	(b) 1009	(c) 502
4. (a) 1202	(b) 496	(c) 909
5. (a) 918 r5	(b) 475	(c) 329 r9

6. Last month: 1380
This month: 1380 x 3 = **4140**
He sold 4140 cakes this month.

7.
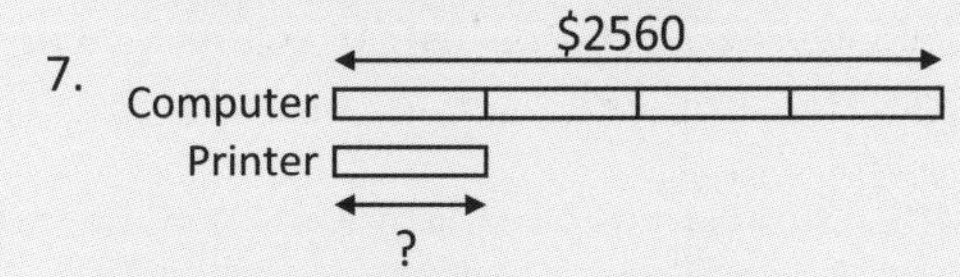

$2560 ÷ 4 = **$640**
The printer costs $640.

8. 1536 ÷ 6 = **256**
Each box had 256 rubber bands.

9. 3750 kg ÷ 10 kg = **375**
He had 375 bags of potatoes.

10.

$9798 – (2 x $3654) = **$2490**
The scooter costs $2490.

11.
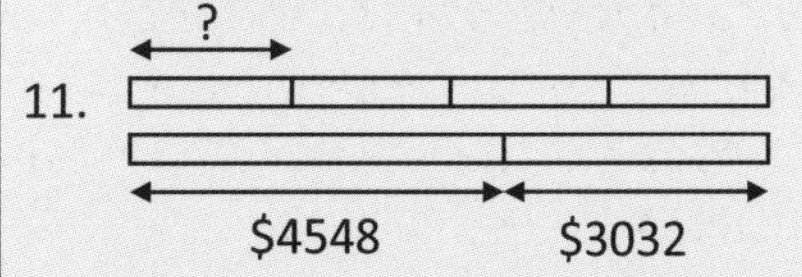

($3032 + $4548) ÷ 4 = **$1895**
He earned $1895 each month.

Chapter 3– Multiplication by a 2-Digit Number

Objectives

- Multiply a multi-digit number by tens.
- Multiply a multi-digit number by tens and ones.
- Multiply a multi-digit number by a 2-digit number.
- Use various mental math strategies to multiply some numbers.
- Estimate the product.

Material

- Place-value discs
- Number cards 1-10, 4 sets
- Mental Math 13-15

Notes

In this chapter your student will learn to multiply by a 2-digit number.

To multiply by a 2-digit number we essentially split the number into tens and ones, multiply by the tens and ones separately, and then add the partial products together.

If your student understands how to multiply by a 1-digit number and understands place-value and the fact that multiplying by tens requires the same process as multiplying by ones, except that the answer is tens, this chapter will be relatively easy. The *Primary Mathematics* curriculum teaches one algorithm that will work in all situations, along with some mental math strategies that can be applied to specific situations. The standard algorithm involves finding the partial products from multiplying by the ones, then by the tens, and then adding the two products together. Your student can either multiply by the ones and then by the tens, or by the tens and then by the ones. Multiplying by the ones first will help him align the digits and remember to write the 0 when multiplying by tens.

Encourage your student to estimate answers. A common error is forgetting to put the 0 in the ones place when multiplying by the tens. Finding an estimate allows her to check for such an error. 67 x 54 ≈ 70 x 50 = 3500. 603 is not a reasonable answer.

Your student can use various mental math strategies in special cases. For example, if one of the factors is close to a multiple of ten or 100, we can multiply by that multiple and subtract the excess. Sometimes using factors of the numbers simplifies the problem enough to do mentally.

67 x 50 = 67 x 5 tens
67 x 5 = 335,
so
67 x 5 tens = 335 tens

$$\begin{array}{r} 67 \\ \times\ 50 \\ \hline 3350 \end{array}$$

67 x 4 = 268

$$\begin{array}{r} 67 \\ \times\ 4 \\ \hline 268 \end{array}$$

67 x 54 = (67 x 50) + (67 x 4)
= 3350 + 268
= 3618

$$\begin{array}{rr} & 67 \\ & \times\ 54 \\ \hline 67 \times 4 \rightarrow & 268 \\ 67 \times 50 \rightarrow & 3350 \\ \hline & 3618 \end{array}$$

67 x 54
↓ ↓
70 x 50 = 3500

$$\begin{array}{r} 67 \\ \times\ 54 \\ \hline 268 \\ 335 \\ \hline 603 \end{array}$$ X

88 x 99 = 88 x 100 – 88
= 8800 – 88
= 8712

28 x 25 = 7 x 4 x 25
= 7 x 100
= 700

(1) Multiply by tens

Activity

Write the expression **23 x 10** and have your student find the answer. He should already be familiar with multiplying by 10 by simply appending a 0. You can use number discs to illustrate the concept. Each ten and each one can be multiplied by 10 separately. 1 ten x 10 = 10 tens = 1 hundred. 1 one x 10 = 10 ones = 1 ten.

Now write the expression **23 x 30**. 30 is the same as 3 tens, or 3 x 10. We can find the answer by multiplying first by 3 and then 10, or first by 10 and then 3. Each ten becomes 3 hundreds, and each one becomes 3 tens.

So, to multiply a number by a ten, we can first multiply it by the digit in the tens place, and then add a 0.

Point out that we could also first multiply by 10, and then by the digit in the tens place.

23 x 10 = 230

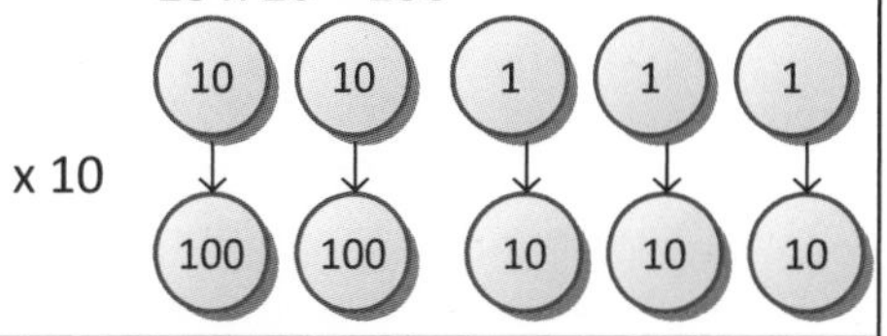

x 10: 10 10 1 1 1 → 100 100 10 10 10

23 x 30

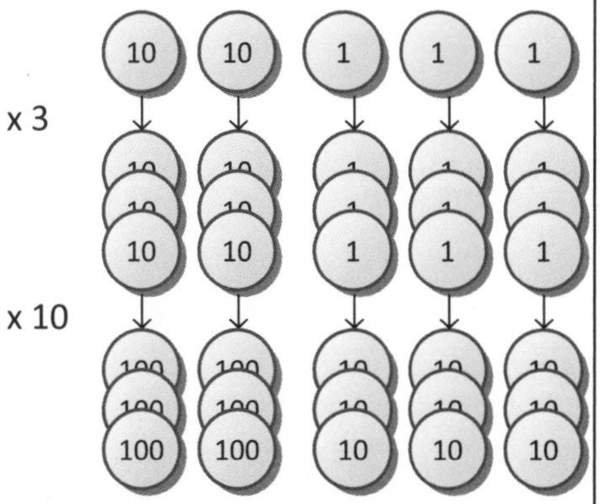

x 3: 10 10 1 1 1 → (10 10 10) (10 10 10) (1 1 1) (1 1 1) (1 1 1)

x 10: → (100 100 100) (100 100 100) (10 10 10) (10 10 10) (10 10 10)

23 x 30 = 23 x 3 x 10
= 69 x 10
= 690

Discussion

Concept p. 68

This illustrates the same concept, except with 32 x 20.

Tasks 1-3, p. 69

1. (a) 160 (b) 400 (c) 2540
(d) 290 (e) 960 (f) 3800
2. 420
420
3. (a) 5680 (b) 12,360

2: The first two methods show that to multiply by 30, we can either first multiply by 10 and then 3, or by 3 and then 10. The third method is similar to the second, but emphasizes that the answer is the same as multiplying by ones and then simply adding a 0.

3: We can rewrite the problem vertically, and multiply in the same way as we would multiply by ones. Show the steps for both multiplying by the ones and by the tens, discussing the place values. For 284 x 2, the little 1 we write above the 2 hundreds to keep track of the answers is a hundred; 80 x 2 is 160, and we write the 1 hundred above and the 6 tens below the line. For 284 x 20, the little 1 is actually a thousand, since we are really multiplying by 2 tens and not 2 ones; 80 x 2 tens = 160 tens = 1600. So it is important that we write the 0 in the answer first to show that the answer is tens. This helps us to place the rest of the digits correctly.

```
 1          1              1
 284        284            284
x  2      x  20   →     x   20
 568          0           5680
```

Practice

Task 4, p. 70

4. (a) 690 (b) 4760 (c) 31,360
(d) 1800 (e) 4050 (f) 33,600

Workbook

Exercise 9, p. 56 (answers p. 81)

(2) Multiply by a 2-digit number

Discussion

Tasks 5-6(b) (Method 1 only), pp. 70-71

5. 510
6. (a) 1728
(b) 5282

If your student thoroughly understands multiplying by tens and by ones, multiplying by a 2-digit number is a simple extension of what she already knows. We simply split the second number into tens and ones, and multiply first by the ones, and then by the tens, and then add the two answers together. By now, the difficulty may be more aligning the digits properly in the correct place than understanding the concept. We write both products under the line, and align them in the correct place so that we can then easily add them together. You can tell her that it does not matter whether we multiply by the ones first or the tens first, but multiplying by the ones first can help us remember to write the 0 for tens down before multiplying by the tens.

Write these problems on paper or whiteboard and guide your student through them. Use place-value terms always. For example, for Task 5: "First we multiply by the ones. 4 ones x 5 ones is 20 ones. Write 2 tens above the next digit and the ones below the line in the ones place. 3 tens x 5 ones = 15 tens, add the 2 tens, that is 17 tens, or 1 hundred 7 tens, write that below the line. Now multiply by the tens. Write a 0 in the ones place. 4 ones x 1 ten = 4 tens. This has no hundred, we can simply write the 4 in the tens place. 3 tens x 1 ten = 3 hundreds. Write the hundreds down. Now add the two numbers..."

Activity

Write the expression **8 x 9** and have your student give the answer. Ask him if he remembers a way to solve this problem if he forgot the math fact for 8 x 9. We can multiply 9 by counting up by 8's (8 x 1 = 8, 8 x 2 = 16...). Since 9 is almost 10, and it is easy to multiply 8 x 10, a faster way is to multiply by 10 and count down by 8, that is, subtract 8. It is easy to subtract from a 10.

Now write the expression **8 x 19** and ask how we can solve this using the same idea. We can count back 8, or subtract 8, from 8 x 20.

Then have your student solve **8 x 49**, **8 x 99**, and **88 x 99** in the same way. For **88 x 99**, we can solve the second step mentally by subtracting 88 from 100.

Have your student solve **888 x 99**. It can be solved in the same way, but is not easy to do mentally. It can still be done using the standard algorithm to subtract. You can have her do it both ways. Point out that alternative methods and mental math strategies work well for some types of problems and save time, particularly problems with smaller numbers, but we can always use the standard algorithm for all kinds of problems.

8 x 9 = 8 x 10 – 8
= 80 – 8
= 72

8 x 19 = 8 x 20 – 8
= 160 – 8
= 152

8 x 49 = 8 x 50 – 8
= 400 – 8
= 392

8 x 99 = 8 x 100 – 8
= 800 – 8
= 792

88 x 99 = 88 x 100 – 88
= 8800 – 88
/ \
8700 100
= 8712

888 x 99 = 888 x 100 – 888
= 88800 – 888
= 87,912

Discussion

Tasks 6(b) (Method 2)-6(d), p. 71

These tasks follow the previous activity. Note that Method 2 for Task 6(b) should not necessarily be done mentally, but Task 6(c) may be done mentally by making 100 with 36.

6. (c) 3564
(d) 700

6(d): This strategy can be used when multiplying by 25 if the other number is a multiple of 4. If your student needs to review the concept of factors, refer to p. 31 in the textbook. (Note: the 2008 printing has an error; the second line of this problem should read 28 x 25, not 28 x 100.)

You may want to point out that since 25 is the same as 100 ÷ 4, we can also multiply by 25 by multiplying by 100 and dividing by 4. We can use this approach even when the other factor is not a multiple of 4, as in 30 x 25, but the division will be harder to do mentally.

28 x 25 = 7 x 4 x 25
= 7 x 100
= 700

28 x 25 = 28 x 100 ÷ 4
= 2800 ÷ 4
= 700

30 x 25 = 30 x 100 ÷ 4
= 3000 ÷ 4
= 750

Practice

Tasks 7-8, p. 72

Have your student use the strategies in Tasks 6(b) and 6(c) to solve these.

7. (a) 792 (b) 2376 (c) 3564
(d) 891 (e) 4653 (f) 3762

8. (a) 200 (b) 900 (c) 1100
(d) 300 (e) 1300 (f) 1800

Workbook

Exercise 10, p. 57 (answers p. 81)

You may want to assign some of Exercise 12 now.

Reinforcement

Give your student some additional problems in order to practice multiplying by a 2-digit number. You can form the numbers randomly using 4 sets of number cards 0-9. Have him copy the problem and then work out the answer.

Mental Math 13

Enrichment

Write some other problems where your student can factor one or more products to make the problem easier to do. Tell her to look for factors that together have a product of 10 or 100. Include some problems that are extensions of multiplying by a number close to a ten.

⇒ 36 x 50 — 36 x 50 = 18 x 2 x 50 = 18 x 100 = 1800

⇒ 18 x 35 — 18 x 35 = 9 x 2 x 5 x 7 = 10 x 63 = 630

⇒ 45 x 12 — 45 x 12 = 9 x 5 x 2 x 6 = 54 x 10 = 540

⇒ 101 x 66 — 101 x 66 = 100 x 66 + 66 = 6666

⇒ 199 x 23 — 199 x 23 = 200 x 23 – 23 = 4600 – 23 = 4577

Mental Math 14

(3) Estimate

Activity

Discuss the problems at the right. In each of them, we can factor out multiples of ten, multiply the remaining factors and the tens or hundreds separately, and then multiply the products. Essentially, we can take off trailing 0's, multiply the non-zero digits, and add the same number of 0's back on to the product.

Try some additional problems such as the following:

⇒ 4**0** x 6**0** = 24**00**

⇒ 4**00** x 6**0** = 24,**000**

⇒ 3**0** x 5**0** = 15**00**

⇒ 3**00** x 5 = 15**00**

⇒ 5 x 8**0** = 40**0**

⇒ 5 x 8**00** = 40**00**

⇒ 5**00** x 8**0** = 40,**000**

3**00** x 1**0** = 3 x 100 x 1 x 10
= 3 x 1 x 100 x 10
= 3 x 1000 = 3**000**

3**00** x 2**0** = 3 x 100 x 2 x 10
= 3 x 2 x 100 x 10
= 6 x 1000
= 6**000**

3**00** x 1**00** = 3 x 100 x 1 x 100
= 3 x 1 x 100 x 100
= 3 x 10,000 = 3**0,000**

3**00** x 2**00** = 3 x 100 x 2 x 100
= 3 x 2 x 100 x 100
= 6 x 10,000
= 6**0,000**

32**0** x 2**00** = 32 x 10 x 2 x 100
= 32 x 2 x 10 x 100
= 64 x 1000
= 64,**000**

Discussion

Tasks 9-10, p. 72

To estimate the product, we round the factors to a number that has one non-zero digit, and then multiply the rounded numbers.

9. (a) 1200 (b) 4000 (c) 6300
   (d) 15,000 (e) 24,000 (f) 20,000
10. 2100
11. 15,000

Remind your student that we want to round to the closest number that makes it easy to use mental math strategies to find the estimated answers. When we want to try to get a close estimate, we can sometimes round to a number with more than one non-zero digit, as in the example at the right; 25 is easy to multiply by 3. If the purpose, though, is in checking answers, we are mostly concerned that we have put things in the correct place value, more than trying to get a closer estimate.

2598 x 307
↓ ↓
2500 x 300 = 750,000 or
3000 x 300 = 900,000
2598 x 307 = 797,586

Practice

Task 12, p. 72

12. (a) 50 x 20 = 1000 882
    (b) 20 x 70 = 1400 1512
    (c) 60 x 50 = 3000 2914
    (d) 400 x 20 = 8000 9476
    (e) 400 x 60 = 24,000 22,214
    (f) 700 x 30 = 21,000 21,920
    (g) 50 x 500 = 25,000 24,990
    (h) 70 x 800 = 56,000 54,234
    (i) 90 x 600 = 54,000 52,272

Workbook

Exercises 11-12, pp. 58-62 (answers pp. 81-82)

You can save some of Exercise 12 for later.

Reinforcement

Extra Practice, Unit 2, Exercise 3, pp. 27-28

Mental Math 15

(4) Practice

Practice

Practice C, p. 73

Tests

Tests, Unit 2, 3A and 3B, pp. 43-46

Enrichment

⇒ Amy has three times as many coins as Zoe. After Amy gave 24 of her coins to Zoe, she had twice as many coins as Zoe. How many coins do they have altogether?

We can draw three units for Amy and one for Zoe, show that 24 move from Amy to Zoe, and that Amy now has 2 units and Zoe 1 unit.

From here, it is not very obvious what to do, since it is difficult to relate the units before to the units after. Encourage your student to try a different approach. Since Amy is giving coins to Zoe, the total amount stays the same. So we could try drawing two bars for each situation, before and after, rather than a bar for each girl. We need 4 units to start with, 3 for Amy and 1 for Zoe. Then we need to decide how to divide the after bar to relate it to the before bar.

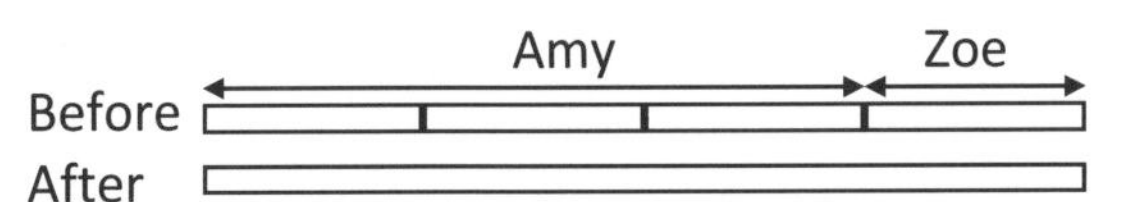

We now want to get equal units for both situations. We need to divide the after bar into three parts, two for Amy and one for Zoe. Ask your student to think of any way he can divide the 4 units into three equal units. One way is to find a common multiple of 4 and 3. The lowest one is 12. So we can divide the bar into 12 equal units. Amy starts with 9 of these smaller units, and Zoe starts with 3. Amy ends up with 8 units, and Zoe with 4. From this, it is easy to see that Amy gave Zoe 1 unit.

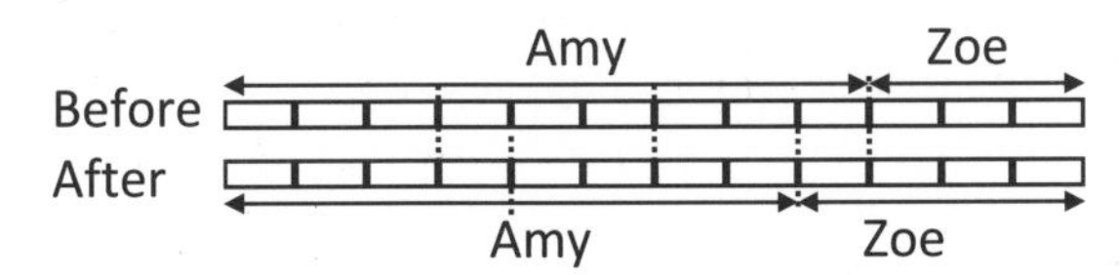

1 unit = 24
12 units = 24 x 12 = 288
They have 288 coins altogether.

---

1. (a) 11,628 (b) 30,160 (c) 55,314
2. (a) 695 (b) 702 (c) 891
3. (a) 528 (b) 1769 (c) 5896
4. (a) 1313 (b) 15,317 (c) 61,308
5. 1 day: 165
   30 days: 165 x 30 = **4950**
   He will deliver 4950 copies in 30 days.
6. 1 sheet: 25 stamps
   15 sheets: 25 x 15 = **375**
   He bought 375 stamps.
7. 1 chair: $128
   12 chairs: $128 x 12 = $1536
   $1536 + $342 = **$1878**
   She started with $1878.
8. 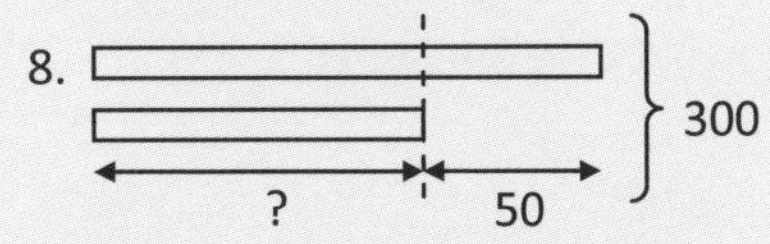

   If 50 is removed from the first group, both would have the same amount.
   2 units = 300 – 50 = 250
   1 unit = 250 ÷ 2 = **125**
   There are 125 children in the second group.
9. 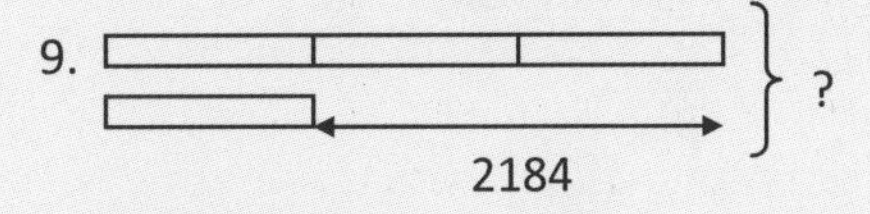

   2 units = 2184
   4 units = 2184 x 2 = **4368**
   The sum of the 2 numbers is 4368.
10. Amount saved in 15 months:
    $65 x 15 = $975
    Amount saved in the next 9 months:
    $2001 - $975 = $1026
    Amount saved each month:
    $1026 ÷ 9 = **$114**
    She saved $114 each of the next 9 months.

Review 2

Review

Review 2, pp. 74-76

14: This involves concepts that have been learned so far, and should be done mentally. You may want to have your student explain how she arrived at her answers.

20: This problem may be challenging if your student has not done earlier levels of *Primary Mathematics*.

Workbook

Review 2, pp. 63-66 (answers p. 82)

Tests

Tests, Units 1-2 Cumulative Tests A and B, pp. 47-53

1. (a) 60 (b) 6000 (c) 600,000

2. (a) 24,038
   (b) 74,002
   (c) 4,300,708
   (d) –100

3. (a) forty-two thousand, three hundred ten
   (b) fifteen thousand, two hundred six
   (c) two hundred eight thousand, one hundred fifty

4. $2500

5. (a) 1, 2 or 4
   (b) 24, 48, 72, 96, 120, ...

6. (a) 3500 (b) 10,000
   (c) 12,464 (d) 10,000

7. (a) 2438 (b) 4025
   (c) 609 (d) 1298

8. (a) 510 (b) 6496
   (c) 1600 (d) 10,000

9. (a) 342 (b) 833 r2
   (c) 712 (d) 314 r5

10. (a) $<$ (b) $>$
    (c) $<$ (d) $<$
    (e) $<$ (f) $<$

11. (a) 2600
    (b) 20,000 (Error in 2008 printing: It should be 1,000,000 + 600,000 + k = 1,620,000)
    (c) 27,000 (d) 1

12. (a) 27 (b) 51
    (c) 8 (d) 4

13. (a) 90,000
    (b) 99,000
    (c) 99,900
    (d) 99,990
    (e) 99,999

14. (a) 8 (b) 4
    (c) 159 (d) 36
    (e) 42 (f) 100
    (g) 72 (h) 6
    (i) 1 (j) 68

15. 475
    271 ?
    (a) (475 – 271) ÷ 3

16. –$48 or $–48

17. –12, –4, –2, 3, 4, 11

18. 3284 ÷ 6 = 547 r2
    (a) There were **547** in each package.
    (b) There were **2** left over.

19. (1240 – 80) ÷ 8 = 1160 ÷ 8 = **145**
    There were 145 children in the group.

20. 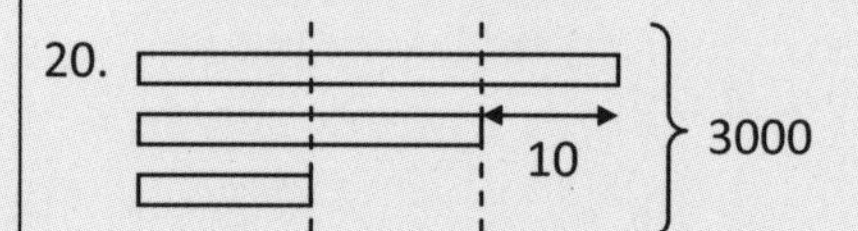

    If 10 books are removed (from the first pile), there will be 5 units.
    5 units = 3000 – 10 = 2990
    1 unit = 2990 ÷ 5 = **598**
    The third pile has 598 books.

Workbook

Exercise 6, pp. 51

1. (a) 2000 x 4 = **8000** 7572
   (b) **4000** x 7 = **28,000** 28,252
   (c) **6000** x 8 = **48,000** 47,896
   (d) **8000** x 9 = **72,000** 73,755

Exercise 7, pp. 52-53

1. (a) 2500 ÷ 5 = **500** 495
   (b) **3600** ÷ 6 = **600** 599
   (c) **4200** ÷ 7 = **600** 602
   (d) **6300** ÷ 9 = **700** 720

2. 12,096 11,850 28,872 43,488
   1302 3069 242 252

Exercise 8, pp. 54-55

1.

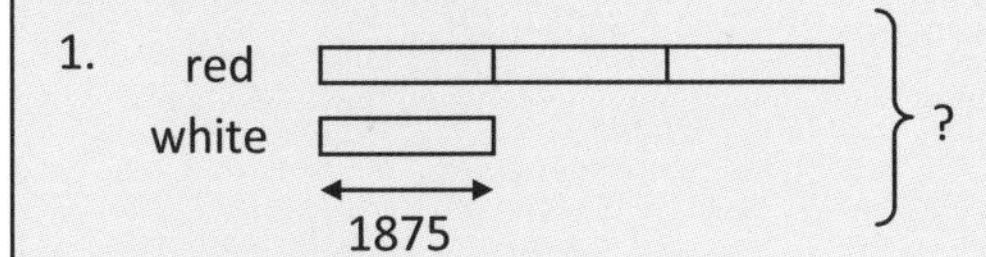

1 unit = 1875
4 units = 1875 x 4 = **7500**
There are 7500 beads.

2.

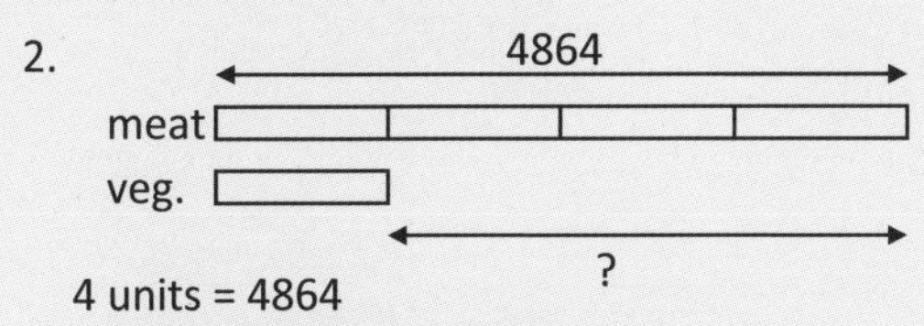

4 units = 4864
1 unit = 4864 ÷ 4 = 1216
3 units = 1216 x 3 = **3648**
There are 3648 more meat than vegetable buns.

3.

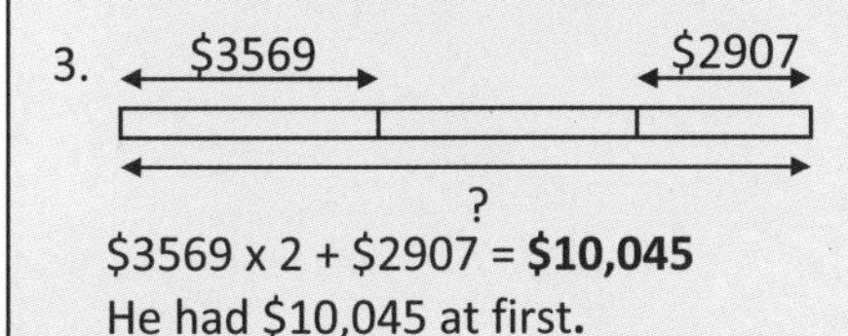

$3569 x 2 + $2907 = **$10,045**
He had $10,045 at first.

4.

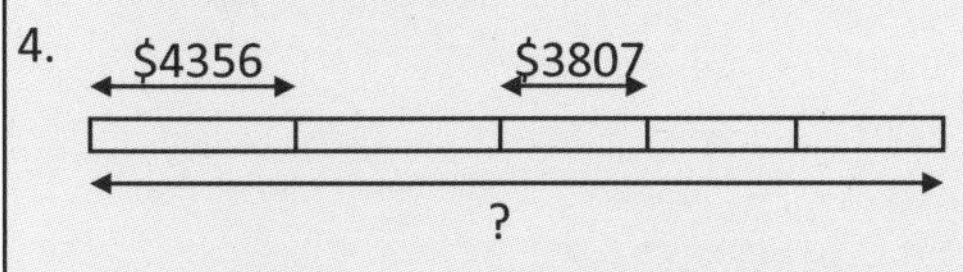

(2 x $4356) + (3 x $3807) = $8712 + $11,421
= **$20,133**
The total money is $20,133.

Exercise 9, p. 56

1. 80 km
   $340
   5860
2. 260 380 5820 7490
   204 2040 200 2000
   1744 17,440 5360 53,600

Exercise 10, p. 57

1. (a) 324
   (b) 5742
   (c) 6900 – 69 = 6831
   (d) 8700 – 87 = 8613

2. (a) 3400
   (b) 800
   (c) 21 x 4 x 25 = 21 x 100 = 2100
   (d) 25 x 4 x 14 = 100 x 14 = 1400

Exercise 11, pp. 58-60

1. (a) 120 120
   1200 1200
   1200 1200
   12,000 12,000
   12,000 12,000

   (b) 30 300
   300
   3000 3000
   3000 3000
   30,000 30,000
   30,000 30,000

2. (a) 2000 2028
   (b) 80 x 30 = 2400 2574
   (c) 30 x 90 = 2700 2523
   (d) 90 x 70 = 6300 6532

3. (a) 8000 8066
   (b) 500 x 60 = 30,000 28,497
   (c) 400 x 60 = 24,000 23,808
   (d) 600 x 80 = 48,000 50,544

Workbook

Exercise 12, pp. 61-62

1. B 273 D 663 F 888 G 6560
   A 868 B 2385 C 3540 E 686

	A 8		B 2	7	C 3	
	D 6	E 6	3		5	
	F 8	8	8		4	
		C 6	5	6	0	

2. Across
   A 2714 C 7719 D 5922

   F 2839 H 1518 J 6225

   Down
   A 27,745 B 41,912 E 9688

   F 21,518 G 3451 I 8775

A 2	7	1	B 4
7			1
C 7	7	1	9
4			1
D 5	E 9	2	2
	6		
	8		
F 2	8	G 3	9
1		4	
5		5	
H 1	5	1	I 8
8			7
			7
J 6	2	2	5

Review 2, pp. 63-66

1. (a) 55,382
   (b) 200,040,012

2. (a) twenty-eight thousand, seven hundred forty
   (b) thirty-five million, eighty-four thousand

3. (a) 38,615; 68,615
   (b) 0, –3, –6

4. (a) 8000
   (b) 10,501
   (c) 67,000

5. 7000

6. (a) 0
   (b) hundred thousands

7. $2312

8. (a) 11,230; 11,290
   (b) –22; –4; 9

9. $90,400

10. 1, 2, 3, 4, 6, 9, 12, 18, 36

11. 23

12. Any 2: 30, 60, 90, 120, 150, ...

13. ___ x 8 = 216 → 216 ÷ 8 = **27**
    The other number is 27.

14. (a) <
    (b) <
    (c) =
    (d) <
    (e) >
    (f) <

15. (a) –1
    (b) –4
    (c) –12
    (d) –5

16. 1 box: 24 apples
    50 boxes: 24 x 50 = 1200
    Groups of 3: 1200 ÷ 3 = 400
    400 groups sell for **$400**.
    He received $400.

17. Adults
    Children
    ?
    9600

    3 units = 9600
    1 unit = 9600 ÷ 3 = **3200**
    There were 3200 children.

18. 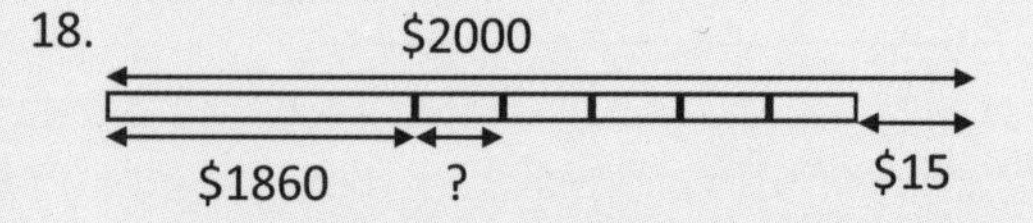

    ($2000 – $1860 – $15) ÷ 5 = $125 ÷ 5 = **$25**
    Each box of CDs costs $25.

Unit 3 – Fractions

Chapter 1 – Equivalent Fractions

Objectives

- Review finding equivalent fractions.
- Review finding the simplest form of a fraction.
- Review comparing and ordering fractions.

Vocabulary

- Equivalent fractions
- Numerator
- Denominator
- Simplest form

Material

- Mental Math 16-18

Notes

In *Primary Mathematics* 2B students learned to understand fractions and write fractional notation, to find sums of fractions that make a whole, and to compare and order unit fractions. In *Primary Mathematics* 3B they learned the terms **numerator** and **denominator** and how to compare and order fractions.

The denominator gives the number of equal parts the whole is divided into. The numerator gives the number of equal parts represented by the fraction. The denominator also indicates the size of the part; the larger the denominator the smaller the size since the whole is divided up into more parts.

$\frac{\text{numerator}}{\text{denominator}} \quad \frac{3}{5}$

When comparing or doing mathematical operations on fractions, each fraction must have the same whole. At this level, students will just be looking at a fraction of one whole. In all problems that involve comparing fractions at this level, students should assume that each fraction being compared is of the same whole.

Fractions of a whole can be illustrated in various ways; concretely with pizzas or cakes or pans of brownies, etc., and pictorially with bars, rectangles, circles, or other shapes divided into equal parts. Because fractions are related to division, and students will later be modeling fraction word problems with models similar to those used for division, fraction bars will be used predominantly in this guide.

$\frac{2}{3}$, $\frac{4}{6}$, $\frac{6}{9}$, and $\frac{8}{12}$ are different names for the same fraction; they all name the same part of the whole. We can change a fraction into an **equivalent fraction** by multiplying or dividing both the numerator and denominator by the same number.

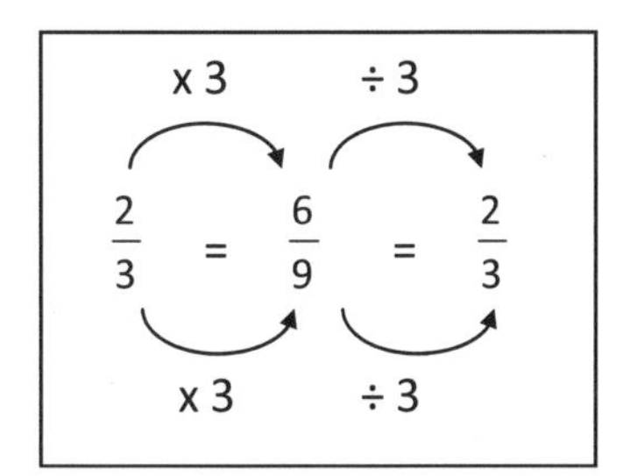

A fraction can be simplified if its numerator and denominator can be divided by the same number. $\frac{6}{9}$ can be simplified to the equivalent fraction $\frac{2}{3}$ by dividing both the numerator and denominator by 3. If it is not possible to divide both the numerator and denominator by any same number (except 1), the fraction is said to be in its **simplest form**. Of the equivalent fractions $\frac{1}{2}$, $\frac{2}{4}$, $\frac{3}{6}$, and $\frac{4}{8}$, $\frac{1}{2}$ is the simplest form. The numerator and denominator do not have a common factor other than 1.

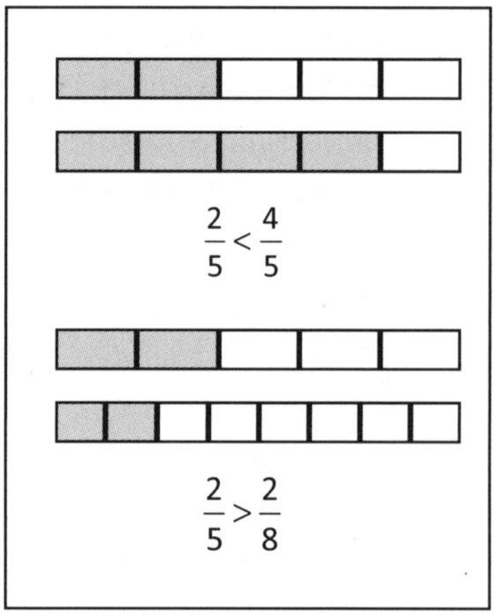

Since the numerator counts the number of parts, it is easy to compare fractions with the same denominator, since the parts are the same size. $\frac{2}{5}$ is smaller than $\frac{4}{5}$. The fraction with the smaller numerator is smaller, since it is the one with the smaller number of parts.

$$\frac{2}{5} < \frac{4}{5}$$

We can also easily compare fractions with the same numerator. $\frac{2}{5}$ is greater than $\frac{2}{8}$ of the same whole because the number of parts is the same, but the size of the parts in $\frac{2}{5}$ is greater than the size of the parts for $\frac{2}{8}$. Therefore, when the numerators are the same, the fraction with the greater denominator is smaller because the size of the parts is smaller.

$$\frac{2}{5} > \frac{2}{8}$$

In order to compare and order fractions where neither the numerator nor the denominator are the same, we need to use equivalent fractions where the numerators or the denominators are the same. In general, students will find equivalent fractions where the denominators are the same, since this will give them practice in finding common denominators when they start adding and subtracting fractions. To compare $\frac{2}{5}$ and $\frac{1}{3}$, we can first find a common multiple of 3 and 5, such as 15. Then we find equivalent fractions with denominators of 15. We can then compare $\frac{6}{15}$ and $\frac{5}{15}$. Another strategy students can use is to find equivalent fractions where the numerators are the same. They could then compare $\frac{4}{10}$ and $\frac{4}{12}$.

$$\frac{2}{5} \downarrow \frac{6}{15} \qquad \frac{1}{3} \downarrow \frac{5}{15}$$

$$\frac{5}{15} < \frac{6}{15} \rightarrow \frac{1}{3} < \frac{2}{5}$$

Unlike fractions can also be compared by multiplying the numerator and denominator of each by the denominator of the other fraction in order to get equivalent fractions where the denominators are the product of the two denominators.

(1) Review: Equivalent Fractions

Discussion

Concept p. 77

$\frac{1}{3}$ $\frac{\text{numerator}}{\text{denominator}}$

Discuss the contents of this page. Remind your student that a fraction tells us the number of parts out of the total number of parts. All the parts are equal size. The first pizza is cut into three equal parts, each a third. We write this as 1 part over 3 total parts. The top number of a fraction is called the *numerator*; it tells us the number of parts we are counting. The bottom number of a fraction is called the *denominator*; it tells us the total number of equal parts in the whole.

$$\frac{1}{3}+\frac{2}{3}=1=\frac{3}{3}$$

The first boy ate one third of the pizza. Ask how much of the pizza is left. One third and two thirds make one whole. The whole is three thirds, which is the same as 1 whole.

If each third is cut again into two equal parts, there are now twice as many parts, each half as large. Each part is now one sixth. The one third now has twice as many parts as well. So two sixths is the same amount of the whole as one third. Similarly, in the third pizza on the page, each of the original thirds is cut into three parts. There are now 9 equal parts, but 3 of these ninths is the same as one of the thirds.

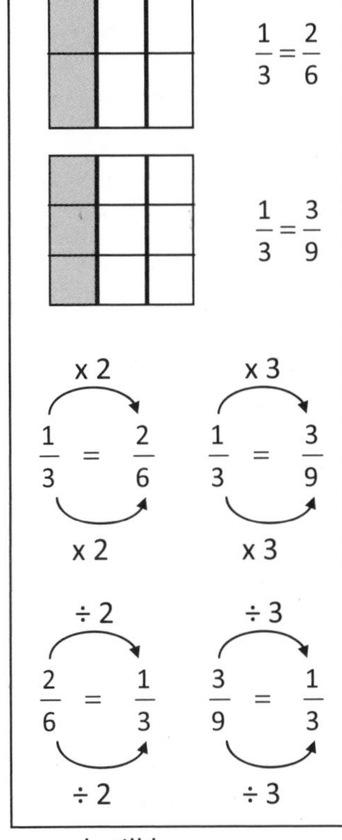

You can show the process with fraction squares or rectangles. Divide a square into thirds with vertical lines and shade one third. Then draw one or two horizontal lines and have your student count the number of shaded parts and the total number of parts in each case. The shaded amount of the total square does not change. Tell him that because they represent the same part of the whole, $\frac{1}{3}$, $\frac{2}{6}$, and $\frac{3}{9}$ are different ways of naming the same fraction. They are *equivalent fractions*. Since we found the equivalent fractions by doubling or tripling both the number of shaded parts and the number of total parts, we can find equivalent fractions by multiplying both the numerator and denominator by the same number.

Ask your student to imagine that we are combining or joining the pieces back together. You can erase the horizontal lines on the fraction squares to illustrate this. In going from two sixths to one third, two parts become one part. Both the shaded number of parts and the total number of parts are halved, or divided by 2. In going from three ninths to one third, three parts become one part. Both the shaded number of parts and total number of parts are divided by 3. If both the numerator and denominator can be divided by the same number, then the resulting fraction is an equivalent fraction.

Point out that once we have thirds, we cannot combine parts any more and still have an equivalent fraction. One third is called the *simplest form* of the fraction; it is the form with the fewest possible number of parts.

Activity

Ask your student to find the simplest form for $\frac{24}{36}$. We need to find a number that both the numerator and the denominator can be divided by; that is, we need to find a common factor of both the numerator and denominator. The easiest ones to start with are 2 or 5. If they both end with 0 or 5, both have 5 as a common factor. If they are both even, both have 2 as a factor. In this case, they are both even, so we can keep dividing by two until they are no longer even. Then we can try 3, and then other prime numbers. We can also find the simplest form in one step; since 12 is the greatest common factor of 24 and 36 we could also have simply divided both numerator and denominator by 12.

$$\frac{24^{\div 2}}{36_{\div 2}} = \frac{12^{\div 2}}{18_{\div 2}} = \frac{6^{\div 3}}{9_{\div 3}} = \frac{2}{3}$$

$$\frac{24^{\div 12}}{36_{\div 12}} = \frac{2}{3}$$

Discussion

Task 1, p. 78

Write the answers as your student finds them, or have her write the answers. Point out that in order to find the missing numerator for $\frac{2}{3} = \frac{\square}{6}$, we have to determine what 3 must be multiplied by to get 6, and then multiply 2 by the same number. Also point out that $\frac{2}{6}$ and $\frac{3}{9}$ are equivalent fractions even though there is no whole number that we can multiply both the numerator and denominator of $\frac{2}{6}$ by to get $\frac{3}{9}$. They are both equivalent to $\frac{1}{3}$ so they are equivalent to each other.

1. (a) $\frac{2}{3} = \frac{\boxed{4}}{6} = \frac{6}{\boxed{9}} = \frac{\boxed{8}}{\boxed{12}}$

(b) $\frac{3}{4} = \frac{6}{\boxed{8}} = \frac{\boxed{9}}{12} = \frac{\boxed{12}}{\boxed{16}}$

Ask your student which of the fractions are in simplest form. $\frac{2}{3}$ and $\frac{3}{4}$ are in simplest form. For both of them, there is no common factor of both the numerator and the denominator (other than 1).

Practice

Tasks 2-4, p. 78

Workbook

Exercise 1, pp. 67-68 (answers p. 90)

Enrichment

Write 6 or more numbers between 1 and 12 and have your student make as many fractions in simplest form as he can with them. For example, if the numbers are 12, 4, 5, 3, 8, 9, some fractions in simplest form that can be formed from them are $\frac{4}{5}$, $\frac{4}{9}$, $\frac{5}{12}$, $\frac{5}{8}$, $\frac{5}{9}$, $\frac{3}{4}$, $\frac{3}{5}$, $\frac{3}{8}$, and $\frac{8}{9}$.

2. (a) $\frac{4}{5} = \frac{\boxed{8}}{10}$ (b) $\frac{1}{4} = \frac{3}{\boxed{12}}$

(c) $\frac{1}{6} = \frac{\boxed{4}}{24}$ (d) $\frac{2}{3} = \frac{10}{\boxed{15}}$

3. (a) $\frac{8}{12} = \frac{\boxed{4}}{6}$ (b) $\frac{9}{15} = \frac{3}{\boxed{5}}$

(c) $\frac{12}{16} = \frac{\boxed{3}}{4}$ (d) $\frac{5}{20} = \frac{1}{\boxed{4}}$

4. (a) $\frac{1}{2}$ (b) $\frac{2}{3}$

(c) $\frac{5}{6}$ (d) $\frac{4}{5}$

(2) Review: Compare and order fractions

Activity

Write down the following sets of fractions and ask your student to tell you which one is smaller. If necessary, discuss how we can determine which one is smaller, and illustrate with fraction bars.

⇒ $\frac{2}{5}$ and $\frac{3}{5}$: The two fractions have the same denominator, so the size of the parts is the same. Two fifths is smaller than three fifths because there are fewer parts.

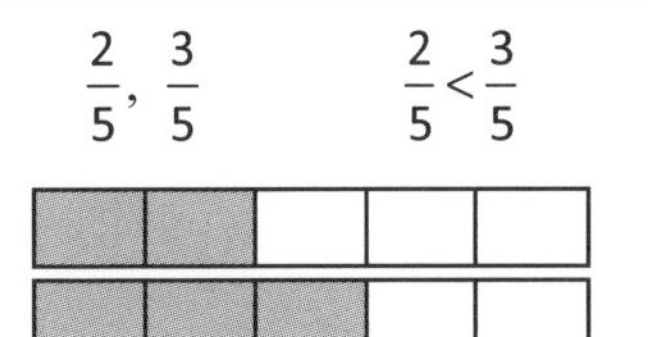

⇒ $\frac{3}{5}$ and $\frac{3}{6}$: Since the denominators are not the same, the size of the parts is not the same. Since the numerators are the same, the number of parts in each fraction are the same. The fraction with the smaller parts, which is the one with the greater denominator, will therefore be smaller.

⇒ $\frac{5}{8}$ and $\frac{3}{4}$: To compare fractions, they either need the same numerator or the same denominator. Since the denominator 8 is a multiple of the denominator 4, we only need to find the equivalent fraction for three fourths with a denominator of 8 in order to compare them.

$\frac{5}{8}, \frac{3}{4} \rightarrow \frac{6}{8}$ $\qquad$ $\frac{5}{8} < \frac{3}{4}$

⇒ $\frac{3}{5}$ and $\frac{2}{3}$: To compare these, we need to find equivalent fractions of both until we get ones with the same denominator or numerator. We can start with the fraction with the largest denominator, three fifths, and list equivalent fractions until we get one that has a denominator that we recognize to be a multiple of the denominator of the smaller fraction, in this case 15. Then we can find the equivalent fraction of two thirds with a denominator of 15 and can compare the fractions.

We could also simply find a common multiple of 5 and 3, and then the equivalent fractions with the common multiple as the denominator. 15 is a common multiple of 5 and 3.

$\frac{3}{5}, \frac{2}{3}$

$\frac{3}{5}, \frac{6}{10}, \frac{9}{15}$

$\frac{2}{3} \rightarrow \frac{10}{15}$ $\qquad$ $\frac{3}{5} < \frac{2}{3}$

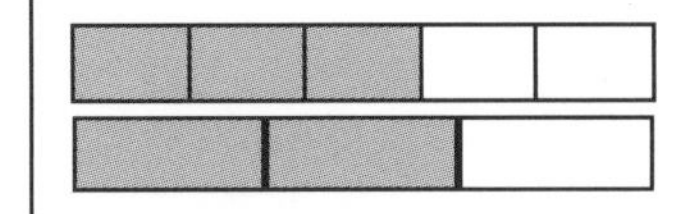

If necessary, you can show the process visually using fraction squares. Divide the square vertically into fifths (using four lines) and horizontally into thirds (using two lines). Color three fifths one way and two thirds another way. Your student can see that $\frac{9}{15}$ squares are colored one way, and $\frac{10}{15}$ squares the other way. After the overlap, there is one fewer square colored for $\frac{3}{5}$ than for $\frac{2}{3}$, so $\frac{3}{5}$ is smaller.

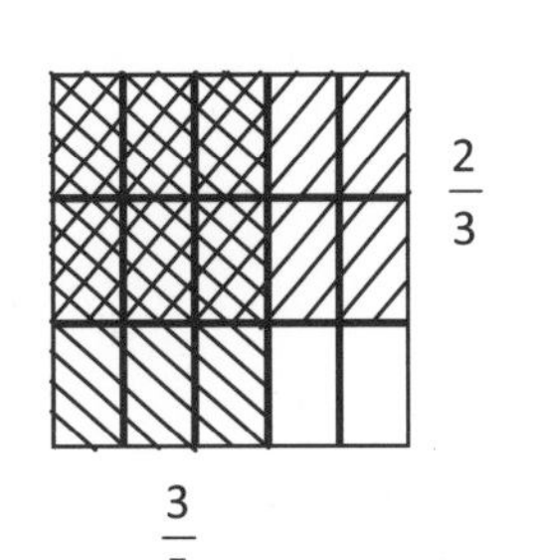

Discussion

Tasks 5-7, p. 79

5(a): To compare the three fractions, we need to find equivalent fractions for all 3 that have the same denominator. Since 10 is a multiple of both 5 and 2, we can use that.

You may also want to show your student that we can compare the three fractions by finding equivalent fractions with the same numerator. 9 is a multiple of both 3 and 1, so we could compare equivalent fractions with 9 in the numerator.

5(b): Ask your student to first find the lowest common multiple of 2, 4, and 3. (12)

After Task 7, remind your student that even though a fraction is written as one number on top of the other, it is a *single* number. The top is the number of parts, the bottom the size of the parts. A fraction can be represented on a number line by dividing the distance between 0 and 1 into parts. The fraction represents a single point on the number line, just as a whole number does, and so it is a single number. Point out that $\frac{2}{3}$ is the same place on the number line as $\frac{4}{6}$. They are the same number.

5. (a) (alternate method)

$$\frac{3}{5} \quad \frac{1}{2} \quad \frac{9}{10}$$
$$\downarrow \quad \downarrow \quad \downarrow$$
$$\frac{9}{15} \quad \frac{9}{18} \quad \frac{9}{10}$$
$$\frac{9}{18} < \frac{9}{15} < \frac{9}{10} \rightarrow \frac{1}{2} < \frac{3}{5} < \frac{9}{10}$$

(b) $\frac{1^{\times 6}}{2_{\times 6}} = \frac{6}{12}$ $\quad \frac{3^{\times 3}}{4_{\times 3}} = \frac{9}{12}$ $\quad \frac{2^{\times 4}}{3_{\times 4}} = \frac{8}{12}$

$\frac{1}{2}, \frac{2}{3}, \frac{3}{4}$

6. (a) 4

(b) $\frac{2}{3}$

7. (a) A: $\frac{2}{3}$

(b) B: $\frac{2}{6}$ C: $\frac{3}{6}$ D: $\frac{5}{6}$

Workbook

Exercise 2, pp. 69-70 (answers p. 90)

Reinforcement

Extra Practice, Unit 3, Exercise 1, pp. 35-36

(3) Practice

Practice

Practice A, p. 80

Tests

Tests, Unit 3, 1A and 1B, pp. 55-59

Enrichment

Have your student look at 1(a) and 1(b). Ask him how he could determine which one is larger without having to find equivalent fractions. For all the equivalent fractions of $\frac{1}{2}$, the numerator is half the denominator. You can have your student list some equivalent fractions of one half to verify this. 3 is more than half of 5, so $\frac{3}{5}$ is more than $\frac{1}{2}$. 3 is less than half of 7, so $\frac{3}{7}$ is less than $\frac{1}{2}$.

Now ask your student to put the fractions in 4(e) in order by comparing each to one half. One fraction is obviously less than one half, and one is greater than one half, so these could be put in order without finding any equivalent fractions.

You can also tell your student that we don't always have to use the lowest common multiple in finding equivalent fractions. For example, in 1(e), the lowest common multiple of the denominators is 18. But another common multiple is the product of the two denominators. All we need to do is multiply each numerator by the denominator of the other fraction to determine which of the two fractions is smallest.

1. (a) $\frac{3}{5} > \frac{1}{2}$ (b) $\frac{3}{7} < \frac{1}{2}$ (c) $\frac{12}{12} = 1$
   (d) $\frac{6}{8} = \frac{9}{12}$ (e) $\frac{5}{6} > \frac{7}{9}$ (f) $\frac{3}{4} > \frac{7}{10}$

2. (a) $\frac{2}{7}$ (b) $\frac{7}{9}$
   (c) $\frac{3}{10}$ (d) $\frac{5}{12}$

3. (a) $\frac{1}{2}$ (b) $\frac{3}{5}$ (c) $\frac{2}{3}$ (d) $\frac{5}{6}$
   (e) $\frac{2}{3}$ (f) $\frac{3}{4}$ (g) $\frac{1}{2}$ (h) $\frac{1}{5}$

4. (a) $\frac{4}{8}, \frac{5}{8}, \frac{7}{8}$ (b) $\frac{6}{12}, \frac{9}{12}, \frac{10}{12}$
   (c) $\frac{1}{8}, \frac{1}{4}, \frac{1}{2}$ (d) $\frac{1}{5}, \frac{3}{10}, \frac{4}{5}$
   (e) $\frac{2}{5}, \frac{1}{2}, \frac{7}{10}$ (f) $\frac{5}{8}, \frac{2}{3}, \frac{3}{4}$

5. (a) $\frac{6}{10}$ ℓ or $\frac{3}{5}$ ℓ
   (b) $\frac{2}{5}$ ℓ

6. **Brian** jogged a longer distance.

$$\frac{5}{6}^{\times 9=45} > \frac{7}{9}^{\times 6=42} \quad \left(\frac{45}{54} > \frac{42}{54}\right)$$

Workbook

Exercise 1, pp. 67-68

1. (a) $\frac{\mathbf{4}}{8}$ (b) $\frac{\mathbf{3}}{5}$

2. (a) $\frac{6}{\mathbf{9}}$ (b) $\frac{3}{\mathbf{4}}$

3. (a) $\frac{3}{12}$ (b) $\frac{8}{10}$

   (c) $\frac{1}{2}$ (d) $\frac{3}{4}$

4.

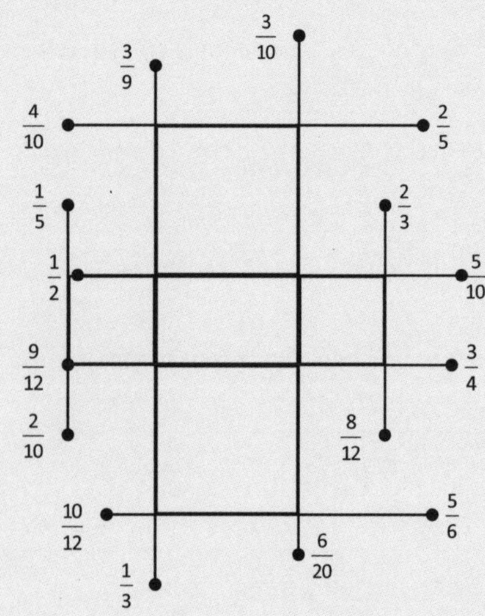

Exercise 2, pp. 69-70

1. (a) $\frac{1}{3}$ $\frac{2}{6}$ $\frac{3}{\mathbf{9}}$ $\frac{\mathbf{4}}{12}$

   (b) $\frac{5}{10}$ $\frac{\mathbf{3}}{6}$ $\frac{4}{\mathbf{8}}$ (c) $\frac{\mathbf{3}}{12}$ $\frac{4}{\mathbf{16}}$ $\frac{1}{4}$

   (d) $\frac{3}{\mathbf{4}}$ $\frac{6}{8}$ $\frac{\mathbf{9}}{12}$ (e) $\frac{12}{\mathbf{12}}$ $\frac{10}{10}$ $\frac{\mathbf{8}}{8}$

2. (a) $\frac{\mathbf{8}}{12}$ (b) $\frac{\mathbf{8}}{10}$

   (c) $\frac{2}{\mathbf{3}}$ (d) $\frac{5}{\mathbf{6}}$

3. (a) $\frac{3}{4}$ (b) $\frac{4}{5}$

   (c) $\frac{2}{3}$ (d) $\frac{1}{2}$

   (e) $\frac{2}{5}$ (f) $\frac{3}{4}$

   (g) $\frac{2}{5}$ (h) $\frac{2}{5}$

4. (a) $\frac{3}{5}$; $\frac{4}{5}$

   (b) $\frac{3}{8}$; $\frac{6}{8}$; $\frac{7}{8}$

5. (a) $\frac{3}{10}$, $\frac{2}{5}$, $\frac{7}{10}$

   (b) $\frac{5}{12}$, $\frac{1}{2}$, $\frac{3}{4}$

Chapter 2 – Adding and Subtracting Fractions

Objectives

- Add like or related fractions.
- Subtract like or related fractions.
- Solve simple word problems involving addition and subtraction of fractions.

Notes

In *Primary Mathematics* 3B, students learned how to add and subtract fractions with like denominators. This is briefly reviewed in this chapter. Your student will learn to add or subtract related fractions; that is, fractions where one denominator is a simple multiple of the other(s). For example $\frac{3}{4}$ and $\frac{3}{8}$ are related fractions. In *Primary Mathematics* 5A, students will learn to add unrelated fractions; that is, fractions where one denominator is not a simple multiple of the other. For example $\frac{2}{7}$ and $\frac{3}{5}$ are unrelated fractions.
In this chapter, the sum will always be less than or equal to a whole. In order to add or subtract fractions, whether related or unrelated, the size of the fractional units have to be the same, so the fractions may have to be renamed to equivalent fractions such that they all have the same denominator.

Since students will be using fraction bars later in solving word problems involving fraction of a set, fraction bars are used the most in this curriculum to illustrate the concepts. The illustration on p. 83 of the textbook shows the addition of $\frac{2}{3}$ and $\frac{1}{6}$ with fraction bars. Do not require your student to draw bars for these problems; he would have to know before drawing the bar that sixths are half of thirds, by which time he might as well have simply found the equivalent fractions. The importance of the pictorial representation is to emphasize that the parts, or fractional units, need to be the same size in order to add them together.

If your student thoroughly understands equivalent fractions from the previous chapter then further pictorial representation is not needed. In some places you might see use of an area model for addition and subtraction of fractions such as that shown at the right. However, even the process of drawing fraction squares and knowing how to divide the square the second time with one line, or divide into sixths vertically and then determine how many new squares need to be shaded, requires an underlying understanding of equivalent fractions to begin with. This type of area model will be used with multiplication of fractions, and to use it now might confuse your student later in distinguishing what to do with the denominator when adding fractions versus when multiplying fractions. This process is therefore not used in *Primary Mathematics*. If your student is struggling with adding fractions, it is better to return to basic concepts of equivalent fractions than add a new type of model at this point.

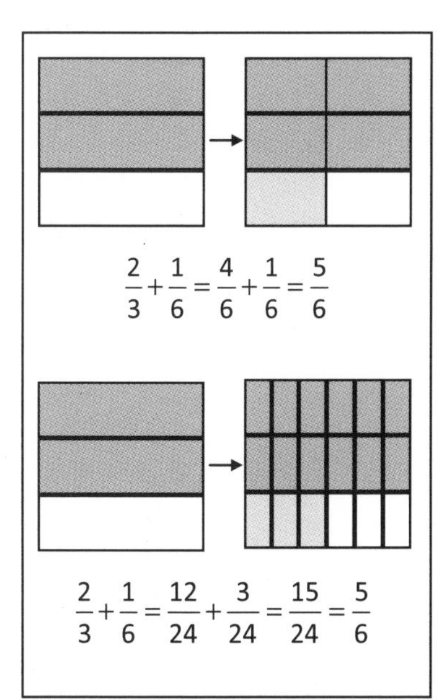

(1) Add related fractions

Discussion

Concept pages 81-82

If necessary, you can draw circles on the board to illustrate this problem as you discuss it. When we add fractions, we are adding parts of a whole. In order to add the parts, they need to be the same size. We can easily find the fraction of the pizza Chrissy and Sharon ate. Since those fractions have the same denominator, the size of the parts are the same. Once we find the total fraction the two girls ate, we simplify the answer to one half. Simplifying the answer gives us a better picture of the fraction of the pizza that has been eaten; it is easier to visualize one half than four eighths.

$\frac{3}{8}+\frac{1}{8}=\frac{4}{8}=\mathbf{\frac{1}{2}}$

$\frac{3}{8}+\frac{1}{8}+\frac{1}{4}=\frac{3}{8}+\frac{1}{8}+\frac{2}{8}=\mathbf{\frac{6}{8}}=\mathbf{\frac{3}{4}}$

$\frac{1}{2}+\frac{1}{4}=\mathbf{\frac{2}{4}}+\frac{1}{4}=\mathbf{\frac{3}{4}}$

To add the three fractions representing how much all three children ate, the fractions have to all represent the same size part. So we can use the equivalent fraction two eighths for the fraction that Paul ate. Similarly, to add one half and one fourth, we need to rename them so they both have the same denominator. Since 4 is a common multiple of 2 and 4, we only need to rename one half.

Tasks 1-5, pp 83-84

1: This is a review of adding and subtracting like fractions. Your student can probably do these mentally.

2: To add two thirds and one sixth, we rename two thirds as four sixths. This is illustrated with the fraction bar.

4: This task shows that once we add the fractions together, we can often simplify the answer. Tell your student that he should always simplify the final answer.

1. (a) $\frac{5}{7}$ (b) $\frac{2}{3}$ (c) $\frac{1}{2}$
   (d) $\frac{4}{9}$ (e) $\frac{2}{5}$ (f) $\frac{3}{8}$
   (g) 1 (h) 0 (i) $\frac{2}{3}$

2. $\frac{2}{3}+\frac{1}{6}=\mathbf{\frac{4}{6}}+\frac{1}{6}=\mathbf{\frac{5}{6}}$

3. (a) $\frac{3}{8}+\frac{1}{4}=\frac{3}{8}+\mathbf{\frac{2}{8}}=\mathbf{\frac{5}{8}}$ (b) $\frac{2}{3}+\frac{1}{9}=\mathbf{\frac{6}{9}}+\frac{1}{9}=\mathbf{\frac{7}{9}}$

4. $\frac{1}{5}+\frac{3}{10}=\mathbf{\frac{2}{10}}+\frac{3}{10}=\mathbf{\frac{5}{10}}=\mathbf{\frac{1}{2}}$

5. (a) $\frac{1}{3}+\frac{1}{6}=\mathbf{\frac{2}{6}}+\frac{1}{6}=\mathbf{\frac{3}{6}}=\mathbf{\frac{1}{2}}$ (b) $\frac{1}{2}+\frac{3}{10}=\mathbf{\frac{5}{10}}+\frac{3}{10}=\mathbf{\frac{8}{10}}=\mathbf{\frac{4}{5}}$

Practice

Task 6, p. 84

Mental Math 16

6. (a) $\frac{1}{2}+\frac{1}{8}=\frac{4}{8}+\frac{1}{8}=\mathbf{\frac{5}{8}}$ (b) $\frac{1}{4}+\frac{2}{12}=\frac{3}{12}+\frac{2}{12}=\mathbf{\frac{5}{12}}$ (c) $\frac{2}{3}+\frac{1}{9}=\frac{6}{9}+\frac{1}{9}=\mathbf{\frac{7}{9}}$
   (d) $\frac{1}{2}+\frac{1}{6}=\frac{3}{6}+\frac{1}{6}=\frac{4}{6}=\mathbf{\frac{2}{3}}$ (e) $\frac{2}{5}+\frac{1}{10}=\frac{4}{10}+\frac{1}{10}=\frac{5}{10}=\mathbf{\frac{1}{2}}$ (f) $\frac{2}{3}+\frac{1}{12}=\frac{8}{12}+\frac{1}{12}=\frac{9}{12}=\mathbf{\frac{3}{4}}$
   (g) $\frac{1}{5}+\frac{3}{10}=\frac{2}{10}+\frac{3}{10}=\frac{5}{10}=\mathbf{\frac{1}{2}}$ (h) $\frac{1}{6}+\frac{7}{12}=\frac{2}{12}+\frac{7}{12}=\frac{9}{12}=\mathbf{\frac{3}{4}}$ (i) $\frac{3}{4}+\frac{1}{12}=\frac{9}{12}+\frac{1}{12}=\frac{10}{12}=\mathbf{\frac{5}{6}}$
   (j) $\frac{1}{3}+\frac{1}{9}+\frac{1}{9}=\frac{3}{9}+\frac{1}{9}+\frac{1}{9}=\mathbf{\frac{5}{9}}$ (k) $\frac{1}{2}+\frac{1}{4}+\frac{1}{4}=\frac{2}{4}+\frac{1}{4}+\frac{1}{4}=\frac{4}{4}=\mathbf{1}$ (l) $\frac{1}{4}+\frac{1}{8}+\frac{3}{8}=\frac{2}{8}+\frac{1}{8}+\frac{3}{8}=\frac{6}{8}=\mathbf{\frac{3}{4}}$

Workbook

Exercise 3, pp. 71-72 (answers p. 96)

(2) Subtract related fractions

Discussion

Tasks 7-11, pp. 85-86

With any of these tasks, illustrate the process by actually drawing the fraction circle, bar, or a fraction square, if necessary. Seeing the process, rather than just the finished picture, will help your student's understanding.

7: In this task, a smaller fraction is subtracted from a half. In order to subtract one eighth from a half, we need to rename the half as eighths.

8: In this task, a half is subtracted from a larger fraction. In order to subtract a half, we need to know how many parts that is, so we need to rename the half as eighths.

10: This task illustrates that we need to sometimes simplify the answer.

7. $\frac{1}{2}-\frac{1}{8}=\frac{4}{8}-\frac{1}{8}=\mathbf{\frac{3}{8}}$

8. $\frac{1}{2}=\frac{\boxed{4}}{8}$ $\quad \frac{7}{8}-\frac{1}{2}=\frac{7}{8}-\frac{4}{8}=\mathbf{\frac{3}{8}}$

9. (a) $\frac{3}{4}-\frac{1}{8}=\frac{\mathbf{6}}{8}-\frac{1}{8}=\frac{\mathbf{5}}{8}$ $\quad$ (b) $\frac{7}{10}-\frac{2}{5}=\frac{7}{10}-\frac{4}{10}=\frac{\mathbf{3}}{10}$

10. $\frac{3}{4}=\frac{\boxed{9}}{12}$ $\quad \frac{3}{4}-\frac{5}{12}=\frac{\mathbf{9}}{12}-\frac{5}{12}=\frac{\mathbf{4}}{12}=\mathbf{\frac{1}{3}}$

11. (a) $\frac{7}{10}-\frac{1}{2}=\frac{7}{10}-\frac{\mathbf{5}}{10}=\frac{\mathbf{2}}{10}=\mathbf{\frac{1}{5}}$ $\quad$ (b) $\frac{2}{3}-\frac{5}{12}=\frac{\mathbf{8}}{12}-\frac{5}{12}=\frac{\mathbf{3}}{12}=\mathbf{\frac{1}{4}}$

Practice

Task 12, p. 86

Mental Math 17

12. (a) $\frac{5}{9}-\frac{1}{3}=\frac{5}{9}-\frac{3}{9}=\mathbf{\frac{2}{9}}$ $\quad$ (b) $\frac{3}{4}-\frac{3}{8}=\frac{6}{8}-\frac{3}{8}=\mathbf{\frac{3}{8}}$ $\quad$ (c) $\frac{4}{5}-\frac{7}{10}=\frac{8}{10}-\frac{7}{10}=\mathbf{\frac{1}{10}}$

(d) $\frac{5}{6}-\frac{1}{2}=\frac{5}{6}-\frac{3}{6}=\frac{2}{6}=\mathbf{\frac{1}{3}}$ $\quad$ (e) $\frac{1}{3}-\frac{1}{12}=\frac{4}{12}-\frac{1}{12}=\frac{3}{12}=\mathbf{\frac{1}{4}}$ $\quad$ (f) $\frac{7}{10}-\frac{1}{5}=\frac{7}{10}-\frac{2}{10}=\frac{5}{10}=\mathbf{\frac{1}{2}}$

(g) $\frac{1}{2}-\frac{1}{10}=\frac{5}{10}-\frac{1}{10}=\frac{4}{10}=\mathbf{\frac{2}{5}}$ $\quad$ (h) $\frac{3}{4}-\frac{5}{12}=\frac{9}{12}-\frac{5}{12}=\frac{4}{12}=\mathbf{\frac{1}{3}}$ $\quad$ (i) $\frac{5}{6}-\frac{7}{12}=\frac{10}{12}-\frac{7}{12}=\frac{3}{12}=\mathbf{\frac{1}{4}}$

(j) $1-\frac{1}{2}-\frac{1}{4}=\frac{4}{4}-\frac{2}{4}-\frac{1}{4}=\mathbf{\frac{1}{4}}$ $\quad$ (k) $1-\frac{1}{2}-\frac{1}{6}=\frac{6}{6}-\frac{3}{6}-\frac{1}{6}=\frac{2}{6}=\mathbf{\frac{1}{3}}$ $\quad$ (l) $\frac{2}{3}-\frac{1}{6}-\frac{1}{3}=\frac{4}{6}-\frac{1}{6}-\frac{2}{6}=\mathbf{\frac{1}{6}}$

Workbook

Exercise 4, pp. 73-74 (answers p. 96)

(3) Solve word problems

Activity

Discuss the following problems. Have your student determine whether we need to add or subtract. You can draw bar models if needed. Since the fractions are numbers, you could either draw part-whole or comparison models to determine whether the problem involves addition or subtraction, as you would if the problems involved whole numbers. You could also draw fraction bars and shade the parts involved in the problem to demonstrate the need for equivalent fractions. Which type of model you draw depends on which concept your student has trouble with, because there is no point in finding equivalent fractions until you know that they need to be added or subtracted, and fraction bars will be easier to draw if you already know what equivalent fractions will be needed. Both are shown for the first problem. However, your student probably will not need to draw bars for these problems, since she should both know how to solve simple addition and subtraction problems, and how to add and subtract fractions.

⇒ Harriet likes birds and is often buying new ones. She went to the pet store and spent $\frac{3}{8}$ of her money on a cockatoo and $\frac{1}{4}$ of her money on a parrot. What fraction of her money did she spend?

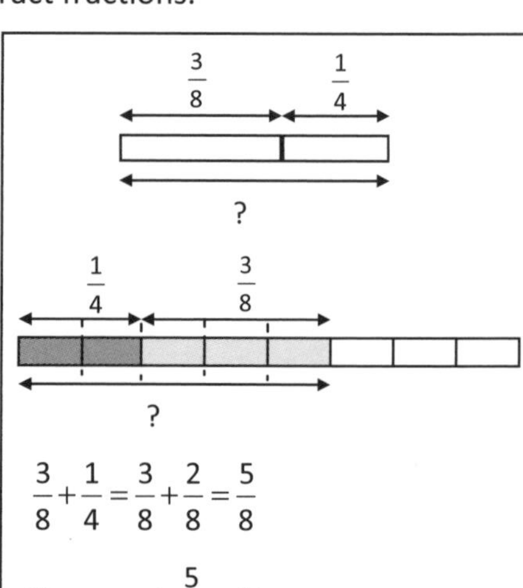

$\frac{3}{8}+\frac{1}{4}=\frac{3}{8}+\frac{2}{8}=\frac{5}{8}$

She spent $\frac{5}{8}$ of her money.

⇒ $\frac{1}{2}$ of Harriet's parakeets are green, $\frac{1}{4}$ of them are yellow, $\frac{1}{8}$ of them are blue, and the rest are white. What fraction of her parakeets are white?

$1-\frac{1}{2}-\frac{1}{4}-\frac{1}{8}=\frac{8}{8}-\frac{4}{8}-\frac{2}{8}-\frac{1}{8}=\frac{1}{8}$

$\frac{1}{8}$ of her parakeets are white.

⇒ Harriet bought $\frac{5}{8}$ lb of birdseed and $\frac{3}{16}$ lb of sprouts for her birds. How much did she buy altogether, in pounds? How much more birdseed did she buy than sprouts?

Point out that when we write $\frac{5}{8}$ lb, that means $\frac{5}{8}$ of a pound.

$\frac{5}{8}\text{ lb}+\frac{3}{16}\text{ lb}=\frac{10}{16}\text{ lb}+\frac{3}{16}\text{ lb}=\frac{13}{16}\text{ lb}$

She bought $\frac{13}{16}$ lb of bird food.

$\frac{5}{8}\text{ lb}-\frac{3}{16}\text{ lb}=\frac{10}{16}\text{ lb}-\frac{3}{18}\text{ lb}=\frac{7}{16}\text{ lb}$

She bought $\frac{7}{16}$ more pounds of birdseed than sprouts.

Practice

Practice B, p. 87

Workbook

Exercise 5, pp. 75-76 (answers p. 96)

Reinforcement

Extra Practice, Unit 3, Exercise 2, pp. 37-42

Tests

Tests, Unit 3, 2A and 2B, pp. 61-65

Enrichment

Mental Math 18

Ask your student the following:

⇒ How many $\frac{1}{8}$'s are in $\frac{3}{4}$? (6)

⇒ Find a fraction that is both a whole and has a denominator of 7. $\left(\frac{7}{7}\right)$

⇒ Find a fraction which, when added to itself, is equal to the sum of $\frac{1}{12}$ and $\frac{1}{4}$. $\left(\frac{1}{6}\right)$

1. (a) $\frac{1}{8}$ (b) $\frac{7}{12}$ (c) $\frac{7}{10}$
2. (a) $\frac{5}{9}$ (b) $\frac{5}{6}$ (c) $\frac{2}{3}$
3. (a) $\frac{1}{12}$ (b) $\frac{1}{8}$ (c) $\frac{1}{2}$
4. (a) $\frac{7}{8}$ (b) $\frac{5}{12}$ (c) 1
5. (a) $\frac{7}{8}$ (b) $\frac{3}{8}$ (c) $\frac{1}{2}$
6. $\frac{3}{4}\ell - \frac{1}{2}\ell = \frac{3}{4}\ell - \frac{2}{4}\ell = \mathbf{\frac{1}{4}}\ell$

   She has $\frac{1}{4}\ell$ of the orange juice left.
7. $\frac{1}{2} + \frac{1}{8} = \frac{4}{8} + \frac{1}{8} = \mathbf{\frac{5}{8}}$

   He used $\frac{5}{8}$ of the paint.
8. (a) $\frac{2}{5}\text{ kg} - \frac{1}{10}\text{ kg} = \frac{4}{10}\text{ kg} - \frac{1}{10}\text{ kg} = \mathbf{\frac{3}{10}\text{ kg}}$

   Courtney bought $\frac{3}{10}$ kg of sugar.

   (b) $\frac{2}{5}\text{ kg} + \frac{3}{10}\text{ kg} = \frac{4}{10}\text{ kg} + \frac{3}{10}\text{ kg} = \mathbf{\frac{7}{10}\text{ kg}}$

   The total weight of sugar is $\frac{7}{10}$ kg.

⇒ Find a fraction that is between $\frac{1}{5}$ and $\frac{1}{3}$, with a numerator that is an odd number less than 10 and a denominator less than 2 tens that is four times the numerator. $\left(\frac{3}{12}\right)$

Workbook

Exercise 3, pp. 71-72

1. (a) $= \frac{\boxed{5}}{\boxed{12}}$

(b) $= \frac{3}{8} + \frac{\boxed{4}}{\boxed{8}} = \frac{\boxed{7}}{\boxed{8}}$

(c) $= \frac{\boxed{4}}{\boxed{10}} + \frac{3}{10} = \frac{\boxed{7}}{\boxed{10}}$

2. $\frac{3}{4} \longrightarrow \frac{5}{6}$

$\frac{8}{9} \longleftarrow \frac{7}{9}$

$\frac{3}{10} \longrightarrow \frac{1}{2}$

$\frac{7}{8} \longleftarrow \frac{5}{8}$

$\frac{3}{4} \longrightarrow \frac{1}{3}$

Exercise 4, pp. 73-74

1. (a) $= \frac{\boxed{1}}{\boxed{4}}$

(b) $= \frac{5}{6} - \frac{\boxed{4}}{\boxed{6}} = \frac{\boxed{1}}{\boxed{6}}$

(c) $= \frac{\boxed{8}}{\boxed{12}} - \frac{1}{12} = \frac{\boxed{7}}{\boxed{12}}$

2. A: $\frac{1}{3}$ D: $\frac{1}{8}$ E: $\frac{4}{9}$

I: $\frac{2}{3}$ L: $\frac{3}{10}$ Q: $\frac{5}{12}$

R: $\frac{1}{2}$ T: $\frac{1}{12}$ U: $\frac{1}{4}$

QUADRILATERAL

Exercise 5, pp. 75-76

1. $1 - \frac{3}{8} = \mathbf{\frac{5}{8}}$

She had $\frac{5}{8}$ of the cloth left.

2. $\frac{3}{4}\text{ m} - \frac{5}{12}\text{ m} = \frac{9}{12}\text{ m} - \frac{5}{12}\text{ m} = \frac{4}{12}\text{ m} = \mathbf{\frac{1}{3}\text{ m}}$

The stick is $\frac{1}{3}$ m longer than the string.

3. $\frac{1}{2} + \frac{1}{6} = \frac{3}{6} + \frac{1}{6} = \frac{4}{6} = \mathbf{\frac{2}{3}}$

He spent $\frac{2}{3}$ of his money.

4. $\frac{3}{10}\ell + (\frac{3}{10}\ell - \frac{1}{5}\ell) = \frac{3}{10}\ell + (\frac{3}{10}\ell - \frac{2}{10}\ell)$

$= \frac{3}{10}\ell + \frac{1}{10}\ell = \frac{4}{10}\ell = \mathbf{\frac{2}{5}\ell}$

They drank $\frac{2}{5}$ ℓ of orange juice altogether.

5. $1\text{ yd} - \frac{1}{2}\text{ yd} - \frac{3}{10}\text{ yd} = \frac{10}{10}\text{ yd} - \frac{5}{10}\text{ yd} - \frac{3}{10}\text{ yd}$

$= \frac{2}{10}\text{ yd} = \mathbf{\frac{1}{5}\text{ yd}}$

She had $\frac{1}{5}$ yd of ribbon left.

Chapter 3 – Mixed Numbers

Objectives

- Interpret mixed numbers as the sum of a whole number and a proper fraction.
- Read and interpret number lines with mixed numbers.
- Add a proper fraction to a whole number.
- Subtract a proper fraction from a whole number.

Vocabulary

- Mixed number
- Proper fraction

Material

- Paper strips, squares, or circles
- Mental Math

Notes

Until now, students have only dealt with fractions less than or equal to 1.

$$2\frac{2}{3} = 2 + \frac{2}{3}$$

A fraction less than 1 is called a **proper fraction**. In this chapter, your student will learn about fractions greater than 1 with a whole number part and a fractional part. These are called **mixed numbers**.

Your student will also learn to subtract a fraction from a whole number greater than 1. Later, when he learns to add and subtract mixed numbers, he can use this skill to "make a whole" or "subtract from a whole" in order to do mental calculations with fractions.

Mixed numbers can be represented by various concrete objects, such as with food (as in two and a half watermelons), with measurements (as in two and a half feet), and pictorially using shapes or fraction bars.

Mixed numbers can also be located on a number line. A mixed number is located at a single point on the number line. Like a fraction, a mixed number represents a single number.

(1) Understand mixed numbers

Discussion

Concept p. 88

Discuss the examples on this page. You can also use concrete objects for examples. Food is easy to use since it can easily be in fractional amounts, such as two and a half bagels or three and a third cups of milk. Make sure your student understands that a mixed number is the sum of a whole number and a fraction. The whole number part is always written first.

Tasks 1-3, p. 89

Have your student actually write the answers to at least Tasks 1 and 3.

For Task 3(d-f), you can illustrate one or more of these with a number bond, showing that we subtract the fraction from one of the wholes.

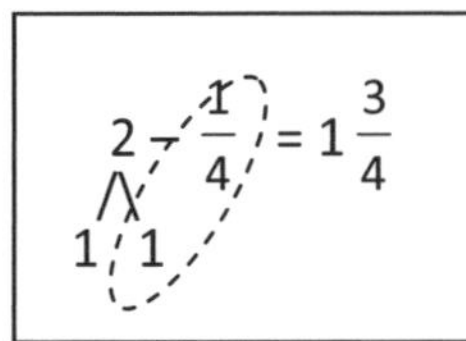

1. (a) $1\frac{1}{3}$ (b) $2\frac{3}{5}$ (c) $2\frac{1}{6}$
2. (a) A: $1\frac{4}{5}$ B: $2\frac{4}{5}$
   (c) C: $1\frac{2}{8}$ D: $1\frac{7}{8}$
3. (a) $3\frac{2}{3}$ (b) $2\frac{4}{5}$ (c) $4\frac{7}{10}$
   (d) $1\frac{3}{4}$ (e) $2\frac{4}{5}$ (f) $4\frac{1}{3}$

Activity

Write a set of mixed numbers and have your student put them in order. Make sure she understands that we can first put them in order by the whole number part. Then, for any two mixed numbers that have the same whole number part, we can put them in order by the fractional part.

$3\frac{3}{8}$ $1\frac{5}{7}$ $2\frac{3}{5}$ $2\frac{3}{4}$

$1\frac{5}{7} < 2\frac{3}{5} < 2\frac{3}{4} < 3\frac{3}{8}$

Workbook

Exercise 6, pp. 77-78 (answers p. 107)

Reinforcement

Extra Practice, Unit 3, Exercise 3, p. 43, problems 1-3

Use a 12-inch ruler. Some rulers show 16ths between the 0 and 6 inch marks and tenths between the 6 and 12 inch marks. Have your student identify what each tick mark on the ruler represents (whole, halves, quarters, fifths, eighths, or sixteenths). Draw some lines and have him measure and write the length as a mixed number. Or ask him to draw some lines of a specified length, giving the length as a mixed number.

Enrichment

Extra Practice, Unit 3, Exercise 3, p. 44, problem 4

4(a) could be solved by imagining each yard cut into 10 equal pieces, and then sticking together one from each yard. There are 28 of them, each a tenth of a yard. 20 of them are two yards, and then there is another eight tenths of a yard left. So each piece would be two and fourth fifths yards. For 4(b), your student could draw a picture.

Tests

Tests, Unit 3, 3A and 3B, pp. 67-71

Chapter 4 – Improper Fractions

Objectives

- Interpret improper fractions as repeated addition of unit fractions.
- Identify improper fractions.
- Convert an improper fraction to a mixed number.
- Convert a mixed number to an improper fraction.

Vocabulary

- Improper fraction

Material

- Paper strips, squares, or circles
- Mental Math 19

Notes

In this chapter your student will learn about improper fractions and how to convert between improper fractions and mixed numbers.

An **improper fraction** is a fraction where the numerator is larger than or equal to the denominator. Improper fractions are equal to or greater than 1. The term fraction means a part of a whole. By that definition, a number that is larger than a whole is not really a fraction. Thus it is an improper fraction; that is, not a proper fraction.

Improper fractions:

$$\frac{7}{4} \quad \frac{9}{9} \quad \frac{26}{25}$$

Your student will learn to convert a mixed number into an improper fraction by first converting the whole part into a fraction and then adding that to the fractional part. Essentially, the numerator of the resulting improper fraction is the whole number multiplied by the denominator plus the numerator of the fractional part.

$$3\frac{1}{5} = \frac{15}{5} + \frac{1}{5} = \frac{16}{5}$$

$$3\frac{1}{5} = \frac{(3\times5)+1}{5} = \frac{16}{5}$$

Your student will also learn to convert an improper fraction to a mixed number by grouping fractions to make whole numbers. Essentially, the numerator is split into a multiple of the denominator and a remainder. In the next chapter, he will relate this process to division.

$$\frac{16}{5} = \frac{15}{5} + \frac{1}{5} = 3\frac{1}{5}$$

$$\frac{16}{5} = \frac{(3\times5)+1}{5} = 3\frac{1}{5}$$

(1) Understand improper fractions

Discussion

Concept p. 90

Explain to your student that if we start with one third, and keep adding thirds, we will end up with more than three thirds, or more than a whole. We can still write a fraction where the numerator is the number of parts, but now the numerator will be greater than the denominator, the number of parts the whole is divided into. An *improper fraction* is a fraction representing a number equal to or greater than 1.

Ask your student how we can easily see whether a fraction is proper or improper. In an improper fraction, the numerator is equal to or larger than the denominator.

The number line shows the relationship between mixed numbers and improper fractions. An improper fraction can be renamed as a mixed number. Make sure your student understands that since both the mixed number and the improper fraction refer to a single point on the number line, they are the same number, just written in two different ways.

$\frac{1}{3}$

$\frac{1}{3}+\frac{1}{3}=\frac{2}{3}$

$\frac{1}{3}+\frac{1}{3}+\frac{1}{3}=\frac{3}{3}=1$

$\frac{1}{3}+\frac{1}{3}+\frac{1}{3}+\frac{1}{3}=\frac{4}{3}=1\frac{1}{3}$

$\frac{1}{3}+\frac{1}{3}+\frac{1}{3}+\frac{1}{3}+\frac{1}{3}=\frac{5}{3}=1\frac{2}{3}$

$\frac{1}{3}+\frac{1}{3}+\frac{1}{3}+\frac{1}{3}+\frac{1}{3}+\frac{1}{3}=\frac{6}{3}=2$

$\frac{1}{3}+\frac{1}{3}+\frac{1}{3}+\frac{1}{3}+\frac{1}{3}+\frac{1}{3}+\frac{1}{3}=\frac{7}{3}=2\frac{1}{3}$

Tasks 1-2, p. 91

Task 1 shows pictorially how to convert a mixed number into an improper fraction, and Task 2 shows pictorially how to convert an improper fraction into a mixed number. For both tasks, ask your student to write both the improper fraction and the mixed number.

Provide more concrete or pictorial examples if your student needs it. Do not ask her to convert between improper fractions and mixed numbers without a pictorial representation yet; that will be covered in the next lessons.

1. There are **7** halves in $3\frac{1}{2}$.

   $\frac{7}{2} = 3\frac{1}{2}$

2. (a) 5 fifths = $\frac{\mathbf{5}}{\mathbf{5}}$ = 1

   (b) 7 quarters = $\frac{\mathbf{7}}{\mathbf{4}} = 1\frac{3}{4}$

   (b) 12 sixths = $\frac{\mathbf{12}}{\mathbf{6}}$ = 2

Workbook

Exercise 7, pp. 79-80 (answers p. 107)

Enrichment

Refer to Task 1. Ask your student the following:

⇒ If there are 7 halves in $3\frac{1}{2}$, how many fourths are there in $3\frac{1}{2}$? (14)

⇒ How many tenths are there in $2\frac{3}{5}$? (26)

(2) Convert an improper fraction to a mixed number

Activity

Use some paper strips, circles, or squares of equal size. Leave one whole for reference and cut four or five of them into fourths.

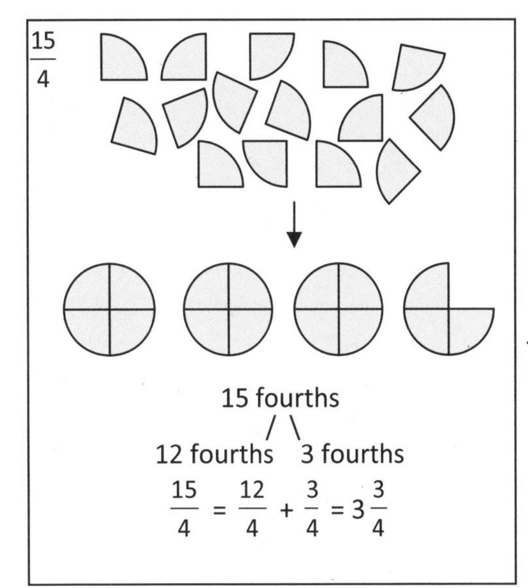

Write the fraction $\frac{15}{4}$. Ask your student to count out 15 of the fourths and arrange them into as many wholes as possible. He should realize that he needs to group the fourths by fours, the denominator of the fraction, to make each whole. Ask him to write the number of fourths as a mixed number. The whole number is the largest multiple of 4 less than 15. The fractional part is the remaining fourths.

To convert an improper fraction into a mixed number, we need to split the improper fraction into two parts, one part containing a multiple of the denominator, and the other the remainder.

Discussion

Tasks 3-4, p. 92

3: Students can count the divisions after the whole number to determine the mixed number.

4: Your student should do this task without a number line or shapes.

3. $1\frac{2}{5}$ $2\frac{4}{5}$

   (a) $1\frac{2}{5}$ (b) $2\frac{4}{5}$

4. $2\frac{1}{6}$

Practice

Task 5, p. 92

5. (a) $4\frac{1}{4}$ (b) $3\frac{1}{3}$ (c) 4 (d) $2\frac{2}{5}$

Workbook

Exercise 8, pp. 81-82 (answers p. 107)

Reinforcement

Write the pairs of mixed numbers and improper fractions shown at the right and ask your student to write $<$, $>$, or $=$ between them.

$5\frac{4}{5}$	$\frac{30}{5}$	($<$)
$\frac{20}{3}$	6	($>$)
$\frac{11}{5}$	$2\frac{1}{5}$	($=$)
$3\frac{3}{4}$	$\frac{7}{2}$	($>$)
$\frac{7}{3}$	$2\frac{1}{7}$	($>$)

(3) Convert a mixed number to an improper fraction

Activity

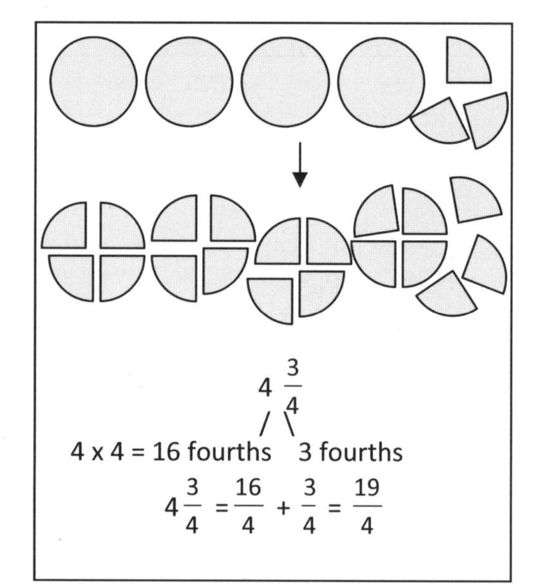

Use some paper strips, circles, or squares of equal size. Cut one of them into fourths and leave 4 of them whole. Give your student the 4 wholes and 3 of the fourths.

Write the mixed number $4\frac{3}{4}$. Ask your student to fold and cut each whole into fourths. She should realize that each whole becomes 4 fourths. Since there were 4 wholes, they become 4 x 4 fourths, or 16 fourths. The total number of fourths is the 16 fourths plus the additional 3 fourths.

To convert a mixed number into an improper fraction, we need to multiply the whole number portion of the mixed number by the denominator of the fractional part, and then add the numerator of the fractional part to get the numerator of the improper fraction.

Discussion

Tasks 6-7, pp. 92-93

6: Your student can add the divisions past the whole number to the numerator of the improper fraction for the whole number.

7: Your student should do this task without a number line or shapes.

6. $\frac{11}{8}$ $\frac{21}{8}$

   (a) $\frac{11}{8}$ (b) $\frac{21}{8}$

7. $\frac{19}{6}$

Practice

Task 8-9, p. 93

9: In this task your student converts only one of the wholes into a fraction and adds it to the fractional part. This strategy will be used in subtracting mixed numbers in *Primary Mathematics* 5. He may notice than an easy way to find the numerator when only one whole is renamed is to add the denominator to it.

8. (a) $\frac{9}{5}$ (b) $\frac{8}{3}$ (c) $\frac{9}{4}$ (d) $\frac{17}{6}$

9. (a) $1\frac{4}{3}$ (b) $1\frac{7}{5}$ (c) $2\frac{5}{4}$

   (d) $2\frac{3}{2}$ (e) $3\frac{7}{6}$ (f) $3\frac{7}{4}$

Workbook

Exercise 9, pp. 83-85 (answers p. 107)

Reinforcement

Extra Practice, Unit 3, Exercise 4, pp. 45-46

(4) Solve problems involving mixed numbers and fractions

Activity

Write the fraction $\mathbf{2\frac{18}{12}}$ and ask your student to simplify it to a mixed number. The fraction is not in simplest form, and is also an improper fraction. It has to be converted to a mixed number and then the whole number part added to the 2. It can be simplified either before or after finding the mixed number. Have her try both methods, and ask her which is easier. Usually it is easier to simplify first because then the calculations involve smaller numbers.

$$2\frac{18}{12} = 2 + 1\frac{6}{12} = 3\frac{1}{2}$$
$$2\frac{18}{12} = 2\frac{3}{2} = 2 + 1\frac{1}{2} = 3\frac{1}{2}$$

Write the expression $\mathbf{\frac{3}{4} + \frac{5}{8}}$ and ask your student to solve it. The first step is to find equivalent fractions. Then we can simply add the numerators. The sum is an improper fraction and needs to be converted to a mixed number. Show him an alternate method similar to the strategies he has learned with mental math. Since the sum of the numerators is larger than the denominator, we know the answer will be greater than 1. We can make a whole by determining how many parts need to be added to one of the fractions to make it a whole, and subtract that from the other fraction.

$$\frac{3}{4} + \frac{5}{8} = \frac{6}{8} + \frac{5}{8} = \frac{11}{8} = 1\frac{3}{8}$$

(2 moved from $\frac{5}{8}$ to $\frac{6}{8}$)

$$\frac{3}{4} + \frac{5}{8} = \frac{6}{8} + \frac{5}{8} = \frac{8}{8} + \frac{3}{8} = 1\frac{3}{8}$$

Write the problem $\mathbf{3 - \frac{2}{3}}$ and discuss the solution. Since the answer should be written as a mixed number in simplest form, the easiest method is to subtract the fraction from one of the wholes. An alternate method is to rename 3 as an improper fraction, subtract, and simplify.

$$3 - \frac{2}{3} = 2\frac{1}{3}$$

(3 split into 2 and 1)

$$3 - \frac{2}{3} = \frac{9}{3} - \frac{2}{3} = \frac{7}{3} = 2\frac{1}{3}$$

Practice

Tasks 10-12, p. 93

If your student needs more examples before doing these independently, discuss some of the problems in these tasks with her and have her do the rest independently.

10. (a) $2\frac{1}{2}$ (b) 4 (c) $2\frac{1}{2}$ (d) $2\frac{2}{3}$
(e) $3\frac{3}{5}$ (f) $4\frac{3}{4}$ (g) $1\frac{3}{4}$ (h) 4

11. (a) $1\frac{2}{3}$ (b) $1\frac{2}{5}$ (c) 1
(d) $1\frac{4}{7}$ (e) $1\frac{1}{2}$ (f) $1\frac{5}{8}$

12. (a) $2\frac{1}{4}$ (b) $1\frac{5}{8}$ (c) $3\frac{1}{2}$
(d) $1\frac{7}{10}$ (e) $1\frac{1}{5}$ (f) $2\frac{2}{7}$

Workbook

Exercise 10, pp. 86-87 (answers p. 108)

Reinforcement

Mental Math 19

Tests

Tests, Unit 3, 4A and 4B, pp. 73-77

Chapter 5 – Fractions and Division

Objectives

- Relate fractions to division.

Material

- Paper strips, squares, or circles
- Mental Math 20

Notes

In this chapter, your student will formally relate fractions to division. He will learn how to express the remainder of a division problem as a fraction. The quotient will now be a single mixed number, rather than a whole number plus a remainder.

Your student may already have an intuitive understanding of the connection between fractions and division. Fractions allow us to answer division problems where we are dividing a smaller number by a larger number, or where the answer is not a whole number. For example, $2 \div 3 = \frac{2}{3}$ or $5 \div 3 = 1\frac{2}{3}$. Your student can now interpret the horizontal line separating the numerator from the denominator as a division symbol; $\frac{2}{3}$ means $2 \div 3$. In algebra, the division symbol will rarely be used, if at all.

Concrete or pictorial examples are used to relate fractions to division. For example, how would you divide 3 equal-sized cakes among 4 children? One method is to divide each cake into fourths. Each child gets a fourth of each cake, and altogether each child will receive $\frac{3}{4}$.

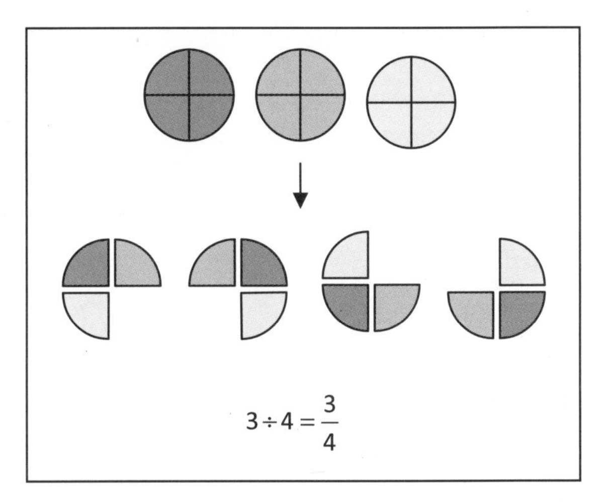

The textbook gives pictorial illustrations for dividing 2 cookies among 3 children on p. 94. The colors show another method than giving each child a third of each cookie; the cookies are divided into thirds, and since there are 6 thirds, we can divide the number of parts by 3: $6 \div 3 = 2$. So each child gets two of the thirds, and in the illustration the first ends up with two thirds of the first cookie, rather than a third of each.

(1) Relate fractions to division

Activity

Use paper circles, squares, or strips to illustrate division of 3 by 4, as in the notes on the preceding page.

Discussion

Concept pp. 94-95

If necessary, use actual paper circles to illustrate these examples concretely.

In the first example on p. 94, there are 6 thirds, so each child gets two thirds.

In the next example, 5 cookies need to be divided among 3 children, and the cookies are again divided into thirds. There are now 15 thirds, so each child gets five of the thirds.

The example on the top of p. 95 shows a different way of dividing 5 by 3; first give each child one of the whole cookies, then divide the remaining 2 into thirds and give each child two thirds. Lead your student to see how this process is similar to the division algorithm. To divide 5 by 3, we can put 1 into each of three groups, so the quotient is 1 and the remainder is 2. But instead of just leaving the remainder as a whole number, we also divide that by 3. $2 \div 3 = \frac{2}{3}$. So now, instead of writing the answer to a division problem that has a remainder as the quotient plus the remainder, we can now give the quotient as a mixed number. The fractional part of the quotient is the remainder over the number we are dividing by.

Tasks 1-2, pp. 95-96

After your student finds the answer to Task 1, you can verify it pictorially by dividing a bar into a unit for each meter. Half of it is clearly three and a half units.

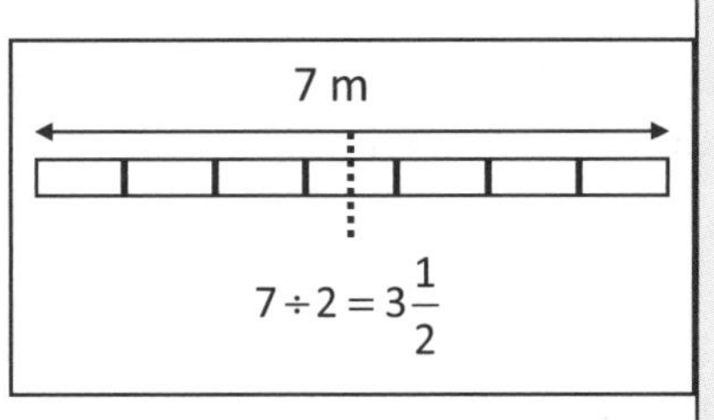

1. $3\frac{1}{2}$; $3\frac{1}{2}$ m

2. $26 \div 8 = 3\frac{2}{8} = 3\frac{1}{4}$

   $26 \div 8 = \frac{26}{8} = \frac{13}{4} = 3\frac{1}{4}$

3. (a) $2\frac{1}{4}$ (b) $2\frac{3}{5}$ (c) $3\frac{1}{3}$

4. $3\frac{3}{4}$

   $3\frac{3}{4}$

Practice

Tasks 3-4, p. 96

Workbook

Exercise 11, pp. 88-89 (answers p. 108)

Reinforcement

Extra Practice, Unit 3, Exercise 5, pp. 47-48, problems 1-4

(2) Practice

Practice

Practice C, p. 97

Tests

Tests, Unit 3, 5A and 5B, pp. 79-82

1. (a) $\frac{7}{5}$ (b) $\frac{10}{9}$ (c) $\frac{8}{3}$ (d) $\frac{20}{9}$
   (e) $\frac{23}{6}$ (f) $\frac{21}{8}$ (g) $\frac{27}{10}$ (h) $\frac{43}{12}$

2. (a) $2\frac{4}{5}$ (b) 5 (c) $1\frac{7}{8}$ (d) $3\frac{1}{2}$
   (e) 3 (f) 6 (g) $2\frac{2}{9}$ (h) $1\frac{1}{2}$
   (i) $3\frac{3}{4}$ (j) $3\frac{3}{4}$ (k) $3\frac{1}{2}$ (l) 3

3. $2 \text{ gal} + \frac{3}{4} \text{ gal} = \mathbf{2\frac{3}{4} \text{ gal}}$

   He got $2\frac{3}{4}$ gallons of gray paint.

4. $2 - \frac{1}{2} = \mathbf{1\frac{1}{2}}$

   She had $1\frac{1}{2}$ cakes left.

5. $3 \text{ lb} \div 6 = \frac{3}{6} \text{ lb} = \mathbf{\frac{1}{2} \text{ lb}}$

   Each share weighed $\frac{1}{2}$ lb.

6. $6 \text{ m} \div 4 = \frac{6}{4} \text{ m} = \mathbf{1\frac{1}{2} \text{ m}}$

   Each piece was $1\frac{1}{2}$ m long.

Enrichment

Your student has learned how to subtract a fraction from a whole number. You can teach her how to subtract a mixed number from a whole number.

Write the expression $\mathbf{3 - \frac{1}{5}}$ and have your student solve it.

Then write the expression $\mathbf{5 - 2\frac{1}{5}}$ and discuss its solution. This is essentially $5 - (2 + \frac{1}{5})$. Both 2 and $\frac{1}{5}$ need to be subtracted from 5. We can first subtract 2, and then subtract $\frac{1}{5}$.

Provide some additional problems. You can use Mental Math 20.

$3 - \frac{1}{5} = 2\frac{4}{5}$ (3 split into 2 and 1)

$5 - 2\frac{1}{5}$

$5 \xrightarrow{-2} 3 \xrightarrow{-\frac{1}{5}} 2\frac{4}{5}$

Workbook

Exercise 6, pp. 77-78

1. (a) $3\frac{1}{2}$
   (b) $2\frac{4}{5}$
   (c) $2\frac{1}{6}$
   (d) $3\frac{7}{8}$
2. (a) B: $1\frac{3}{5}$ m; C: $2\frac{2}{5}$ m
   (b) $3\frac{1}{5}$ ℓ
   (c) $2\frac{3}{4}$
   (d) $2\frac{2}{3}$

Exercise 7, pp. 79-80

1. (a) $\frac{6}{3}$
   (b) $\frac{8}{4}$
   (c) $\frac{11}{6}$
   (d) $\frac{13}{5}$
2. (a) $2\frac{5}{6}$ $\frac{17}{6}$
   (b) $2\frac{4}{9}$ $\frac{22}{9}$
   (c) $1\frac{2}{3}$ $\frac{5}{3}$
   (d) $3\frac{3}{4}$ $\frac{15}{4}$
   (e) $2\frac{3}{5}$ $\frac{13}{5}$
   (f) $2\frac{7}{8}$ $\frac{23}{8}$

Exercise 8, pp. 81-82

1. (a) $2\frac{3}{4}$
   (b) $3\frac{3}{5}$
2. $1\frac{2}{3}$; $2\frac{1}{3}$; 3; $3\frac{2}{3}$
3. (a) $2\frac{1}{2}$ (b) $1\frac{7}{10}$
   (c) $1\frac{1}{6}$ (d) $2\frac{1}{3}$
   (e) $2\frac{1}{5}$ (f) $2\frac{1}{4}$
   (g) $1\frac{3}{8}$ (h) $4\frac{1}{2}$
   (i) 3 (j) 4

Exercise 9, pp. 83-85

1. (a) $\frac{\mathbf{6}}{3}$
   (b) $\frac{\mathbf{6}}{3}$; $\frac{\mathbf{8}}{3}$
2. (a) $\frac{11}{6}$ (b) $\frac{19}{8}$
3. (a) $\frac{7}{5}$ (b) $\frac{5}{4}$
   (c) $\frac{19}{8}$ (d) $\frac{21}{10}$
   (e) $\frac{19}{6}$ (f) $\frac{10}{3}$
   (g) $\frac{5}{2}$ (h) $\frac{23}{5}$
   (i) $\frac{13}{9}$ (j) $\frac{29}{12}$
4. $\frac{4}{4}$; $\frac{7}{4}$; $\frac{9}{4}$; $\frac{11}{4}$; $\frac{14}{4}$
5. $1\frac{1}{9} \leftrightarrow \frac{10}{9}$; $1\frac{1}{8} \leftrightarrow \frac{9}{8}$; $1\frac{1}{7} \leftrightarrow \frac{8}{7}$; $1\frac{1}{6} \leftrightarrow \frac{7}{6}$;
   $1\frac{1}{5} \leftrightarrow \frac{6}{5}$; $1\frac{1}{4} \leftrightarrow \frac{5}{4}$; $1\frac{1}{3} \leftrightarrow \frac{4}{3}$
   $2\frac{2}{3} \leftrightarrow \frac{8}{3}$; $2\frac{1}{2} \leftrightarrow \frac{5}{2}$; $1\frac{3}{4} \leftrightarrow \frac{7}{4}$;
   $2\frac{1}{5} \leftrightarrow \frac{11}{5}$; $1\frac{5}{6} \leftrightarrow \frac{11}{6}$; $1\frac{7}{8} \leftrightarrow \frac{15}{8}$;

Workbook

Exercise 10, pp. 86-87

1. (a) 3 (b) $2\frac{1}{2}$
   (c) 3 (d) $2\frac{2}{3}$
   (e) $6\frac{1}{3}$ (f) $3\frac{1}{3}$

2. $3\frac{7}{4} \leftrightarrow 4\frac{3}{4}$; $3\frac{1}{2} \leftrightarrow 2\frac{3}{2}$; $2\frac{2}{5} \leftrightarrow 1\frac{7}{5}$;
   $2\frac{1}{3} \leftrightarrow 1\frac{4}{3}$; $3\frac{1}{4} \leftrightarrow 2\frac{5}{4}$; $4\frac{1}{6} \leftrightarrow 3\frac{7}{6}$

3. (a) 1 (b) 1
   (c) $1\frac{1}{2}$ (d) $1\frac{2}{7}$
   (e) $1\frac{1}{3}$ (f) $1\frac{1}{8}$
   (g) $1\frac{1}{6}$ (h) $1\frac{1}{10}$

3. (a) $\frac{2}{9}$ (b) $\frac{7}{12}$
   (c) $1\frac{1}{4}$ (d) $1\frac{3}{8}$
   (e) $2\frac{3}{7}$ (f) $\frac{1}{5}$

4. (a) > (b) <
   (c) < (d) =
   (e) < (f) <
   (g) = (h) >
   (i) > (j) =

Exercise 11, pp. 88-89

1. (a) $\frac{3}{2}$
   (b) $\frac{5}{3}$
   (c) $\frac{7}{4}$

2. $2\frac{2}{3}$ $3\frac{1}{3}$ $2\frac{2}{5}$
   $2\frac{3}{4}$ $4\frac{3}{5}$ $6\frac{2}{3}$

3. (a) 4 (b) $2\frac{1}{5}$
   (c) $2\frac{1}{8}$ (d) 9

Chapter 6 – Fraction of a Set

Objectives

- Find the fraction of a set.
- Solve word problems that involve fraction of a set.

Material

- Counters, two colors
- Mental Math 21

Notes

In *Primary Mathematics* 3B, students learned to find the fraction of a set by dividing the set up into equal parts and then finding the amount in the fractional part. In this chapter, your student will review this and learn to interpret a fraction of a set as a fraction times the number in the set. For example, $\frac{3}{5}$ x 20 means $\frac{3}{5}$ of 20.

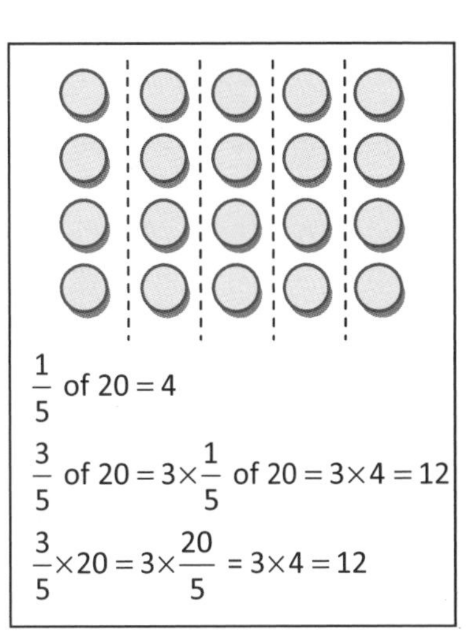

To find $\frac{1}{5}$ of a set of 20 objects, we can divide the set of 20 into 5 equal parts and determine how many objects there are in one part. To find $\frac{3}{5}$ of 20, we can also divide the set of 20 into 5 equal parts. Then we determine how many objects there are in three parts.

From the previous chapters, your student has learned that $\frac{20}{5}$ is the same as 20 divided by 5, that is, putting 20 into 5 equal parts. To find $\frac{1}{5}$ of 20, we are also putting 20 into 5 equal parts. So $\frac{1}{5}$ of 20 is the same as $\frac{20}{5}$, and $\frac{3}{5}$ of 20 is the same as 3 x $\frac{20}{5}$.

In this chapter, your student will also learn to use fraction bars to solve word problems. Each fractional part of the bar is a unit, similar to the unit in the part-whole model for multiplication and division. For example, to find two thirds of 18, we can draw a bar, divide it into 3 units (thirds), divide to find the value of 1 unit, and then multiply to find the value of 2 units.

Bar models are particularly helpful for problems where we are given the value of a fractional part of a whole and need to find the whole. Traditionally, this would be solved by dividing the amount given by the fraction, but students have not learned to do this and do not yet have a conceptual understanding of division of fractions. With the bar models, they can instead solve the problem with units. For example, if we know that $\frac{3}{5}$ of the whole is 15, we can easily find 1 unit by dividing and the whole by multiplying.

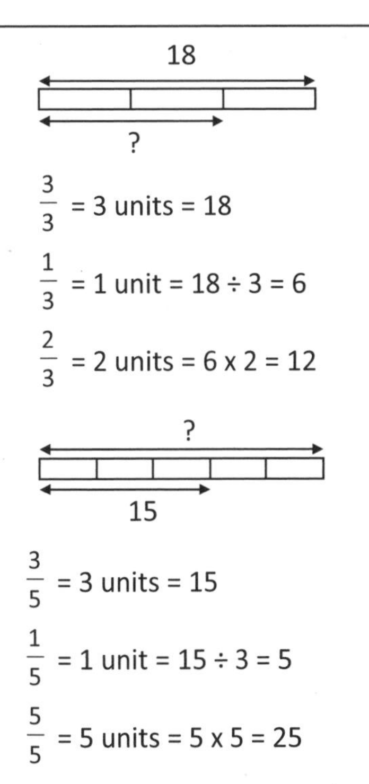

(1) Review: Fraction of a set

Activity

Give your student two counters of one color, such as green, and three of another, such as red. Tell him that up to now, when using fractions, the whole has been one thing, like a shape or a pizza. The whole can also be a set of objects, such as these 5 counters. If we consider each counter to be an equal part of the whole, then we can say that each one is one fifth of the whole. So 2 out of the 5 is two fifths of the 5 counters. Ask him what three fifths of 5 equals.

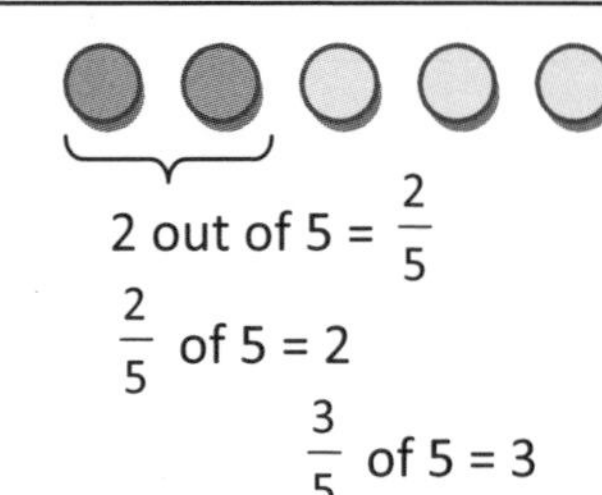

2 out of 5 = $\frac{2}{5}$

$\frac{2}{5}$ of 5 = 2

$\frac{3}{5}$ of 5 = 3

Give your student 6 counters of one color, such as green, and 9 counters of another color, such as red. Ask her to put them into the fewest number of equal groups such that all the counters in a group have the same color. She cannot put all 6 green counters in one group, since if she then puts 6 red counters in another group, there will be 3 left over. The only possible grouping is 3 in each group. Tell her that since we have equal groups, and there are 5 groups, each group of 3 counters is 1 out of 5, or one fifth of the total. Ask her what fraction of the counters are green. There are two groups with green counters, so 2 out of 5 groups, or two fifths, are green. Since there are 6 counters in the two groups, two fifths of 15 is 6. Point out that if we had one counter in each group, then we could say that six fifteenths of the counters are green. Point out that six fifteenths is 6 counters, just like two fifths, and that the fraction six fifteenths is the same as two fifths.

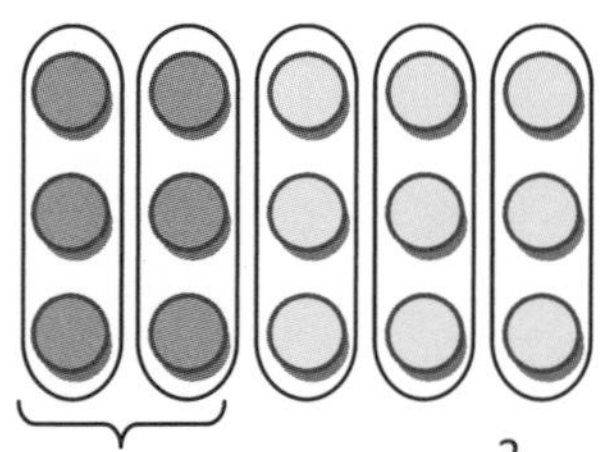

2 groups out of 5 = $\frac{2}{5}$

$\frac{2}{5}$ of 15 = 6

$\frac{6}{15}$ of 15 = 6 = $\frac{2}{5}$ of 15

Give your student 12 counters, six of one color and six of another. Ask him to put them in groups again so that there are only one kind in each group. The largest such group is 6 in each group. He could also make groups of 2 or 3. For each type of group, have him write the fraction of counters that are green. Point out that one half, two fourths, and three sixths of 12 all equal 6 counters, and all the fractions are equivalent fractions.

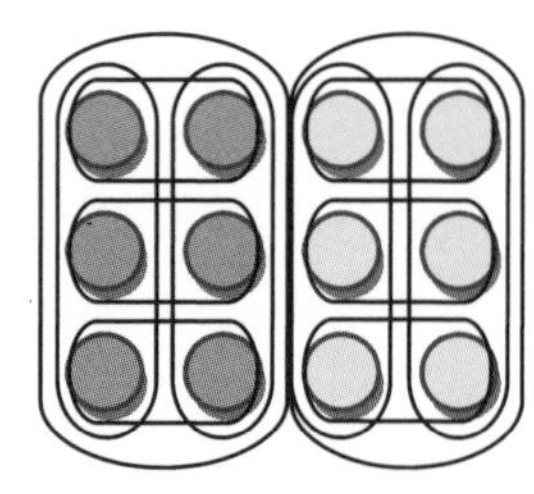

$\frac{1}{2}$ of 12= 6

$\frac{1}{2}$ of 12 = $\frac{2}{4}$ of 12 = $\frac{3}{6}$ of 12

Give your student 20 counters, all of one color. Ask her to use them to find out what one fifth of 20 is. She should group the counters into 5 equal groups. Point out that one fifth of 20 is the same as twenty fifths. By making 5 equal groups, she is dividing 20 by 5. Then ask her to use the answer to one fifth of 20 to find the value of three fifths of 20. Write the equation. Point out that to find the value of different fifths of twenty, we can just multiply the number in the numerator by the value of one fifth of 20.

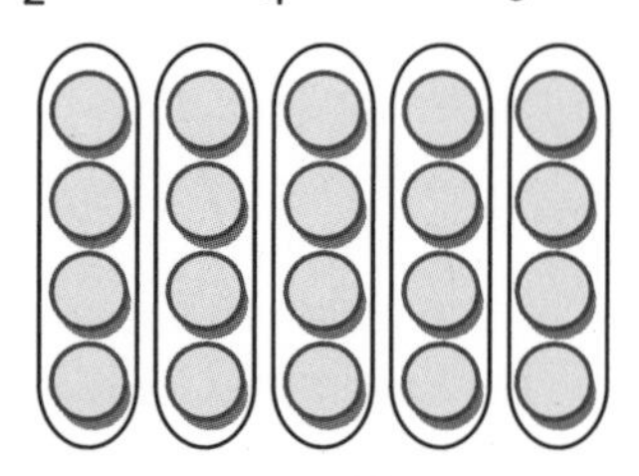

$\frac{1}{5}$ of 20 = 4 = $\frac{20}{5}$

$\frac{3}{5}$ of 20 = 3 x $\frac{1}{5}$ of 20 = 12

Discussion

Concept p. 98

This page covers the same concepts as the first two activities with the counters. If you choose not to use counters, and just skip to the textbook page, be sure you discuss all the concepts in that activity.

Tasks 1-3, p. 99

1. (a) $\frac{1}{2}$

   (b) $\frac{3}{4}$

   (c) $\frac{2}{3}$

   (d) $\frac{5}{6}$

2. $\frac{1}{3}$ of 12 = **4**

3. $\frac{1}{4}$ of 20 = **5**

   $\frac{3}{4}$ of 20 = **15**

Workbook

Exercises 12-13, pp. 90-94 (answers p. 119)

Reinforcement

Give your student 24 counters and have him use them to find the following:

⇒ $\frac{1}{2}$ of 24 (12)

⇒ $\frac{2}{3}$ of 24 (16)

⇒ $\frac{3}{4}$ of 24 (18)

⇒ $\frac{5}{6}$ of 24 (20)

⇒ $\frac{3}{8}$ of 24 (9)

⇒ $\frac{7}{12}$ of 24 (14)

(2) Find the fraction of a set

Discussion

Top of p. 100

Remind your student that on the previous page, the textbook showed that she could find one fourth of 20 counters by dividing the counters into 4 groups and determining how many counters are in one group. This is the same as finding $\frac{20}{4}$. Three fourths of 20 is just 3 of those fourths, so to find three fourths of 20, we can multiply 3 by the answer to $\frac{20}{4}$.

$$\frac{1}{4} \text{ of } 20 = \frac{20}{4} = \mathbf{5}$$

$$\frac{3}{4} \text{ of } 20 = 3\times\frac{20}{4} = 3\times\mathbf{5} = \mathbf{15}$$

Ask your student to find $\frac{1}{5}$ of 8. We would still need to divide 8 into 5 equal groups, which is the same as $\frac{8}{5}$. So we can find the answer by simplifying $\frac{8}{5}$ to a mixed number, or dividing 8 by 5.

$$\frac{1}{5} \text{ of } 8 = \frac{8}{5} = 1\frac{3}{5}$$

Tasks 4-5, p. 100

In these tasks, we are moving away from concrete or pictorial representation of the problem to the abstract. Rather than counting out or drawing 120 counters to find three eighths of 120, we can divide 120 by 8 and then multiply by 3.

4. $\frac{5}{6} \text{ of } 18 = 5\times\frac{18}{6} = 5\times\mathbf{3} = \mathbf{15}$

5. $\frac{1}{8} \text{ of } 120 = \frac{120}{8} = \mathbf{15}$

$\frac{3}{8} \text{ of } 120 = 3\times 15 = \mathbf{45}$

Point out that to find $\frac{120}{8}$, we don't have to just divide 120 by 8. We can do the problem mentally by simplifying the fraction. Since both the numerator and denominator are even numbers, we could divide both by 2, then again by 2, and finally divide by 2. This is easy to do using mental math.

$$\frac{120}{8} \overset{\div 2}{\underset{\div 2}{=}} \frac{60}{4} \overset{\div 2}{\underset{\div 2}{=}} \frac{30}{2} \overset{\div 2}{\underset{\div 2}{=}} \frac{15}{1} = 15$$

Practice

Task 6, p. 100

6. (a) 6 (b) 4 (c) $\frac{2}{3}$
(d) 6 (e) 6 (f) 20
(g) 25 (h) 75 (i) 60

Workbook

Exercise 14, pp. 95-97 (answers p. 119)

Reinforcement

Mental Math 21

(3) Find the fractional part of a whole

Discussion

Tasks 7-9, p. 101

7. $\frac{3}{4}$

8. $\frac{1}{7}$; $\frac{1}{7}$

9. $\frac{4}{25}$

7: To find 6 out of 8, we can put each coin into a group of one. There are 8 equal groups. Each group is $\frac{1}{8}$ of the total set of coins, so 6 of them is $\frac{6}{8}$. But, as we found earlier, we could put the coins into larger groups. If we put 2 coins in each group, we have only pennies in one group, and only dimes in the other 3 groups, with a total of 4 groups. 6 coins is also 3 out of 4 groups, or $\frac{3}{4}$ and is equivalent to $\frac{6}{8}$. So all we have to do to find what fraction of the coins are dimes is put the number of dimes over the total number of coins and simplify the fraction.

9: When we find a measurement as a fraction of a total, both need to be in the same units. So to find 16 cm out of 1 m, we need to convert the meters into centimeters.

To emphasize this point, you can show your student a meter stick. Ask him to locate 25 cm on the meter stick and tell you what fraction of the meter it is. 25 centimeters is one fourth of the way along the meter stick. We do not find what fraction 25 cm is of 1 m by putting 25 over 1, but rather 25 cm over 100 cm. Both units of measurement need to be the same in order to compare them. We convert the larger unit (meters, 1 m) to the smaller unit (centimeters, 100 cm).

25 cm is what fraction of 1 m?

$\frac{25 \text{ cm}}{100 \text{ cm}} = \frac{1}{4}$

25 cm = $\frac{1}{4}$ of 1 m

Activity

Ask your student to find what fraction of a dollar a nickel is.

Since there are 20 nickels in a dollar, then one nickel is $\frac{1}{20}$ of a dollar. We can also convert both the nickel and the dollar to cents and simplify the fraction. Ask her to find what fraction of a dollar 1 penny, 1 dime, and 1 quarter are.

1 nickel is $\frac{5¢}{100¢} = \frac{1}{20}$ of \$1.

1 penny is $\frac{1¢}{100¢} = \frac{1}{100}$ of \$1.

1 dime is $\frac{10¢}{100¢} = \frac{1}{10}$ of \$1.

1 quarter is $\frac{25¢}{100¢} = \frac{1}{4}$ of \$1.

Ask your student to do the following problems.

⇒ What fraction of a gallon is 6 cups?

⇒ 2 months is what fraction of a year?

⇒ What fraction of 3 feet is 6 inches?

$\frac{6 \text{ c}}{16 \text{ c}} = \frac{3}{8}$ — 6 cups is $\frac{3}{8}$ of a gallon.

$\frac{2 \text{ months}}{12 \text{ months}} = \frac{1}{6}$ — 2 months is $\frac{1}{6}$ of a year.

$\frac{6 \text{ in.}}{36 \text{ in.}} = \frac{1}{6}$ — 6 inches is $\frac{1}{6}$ of 3 ft.

Workbook

Exercise 15, pp. 98-99 (answers p. 119)

(4) Solve word problems 1

Discussion

Task 10, p. 102

> 10. 3 units = 24
>
> 1 unit = $24 \div 3 = \frac{24}{3} = \mathbf{8}$
>
> 2 units = $2 \times \frac{24}{3} = 2 \times 8 = \mathbf{16}$
>
> $\frac{2}{3} \times 24 = 2 \times \frac{24}{3} = 2 \times 8 = \mathbf{16}$
>
> There are **16** white flowers.

Make sure your student understands both methods. In Method 1, the whole bar represents the whole, which is 24. Since the problem deals with thirds, we can divide the bar into thirds. Each unit is a third. We need to find two thirds. So we find the value of one unit by dividing 24 by 3, and then the value of 2 units by multiplying that quotient by 2. In Method 2, we don't use bars, but simply realize we are finding two thirds of 24. Tell your student that bars may not be necessary for simpler problems, but it is useful to know how to use them when he has more complex problems.

Point out that $\frac{2}{3}$ of 24 can be written as $\frac{2}{3} \times 24$. When we do a multiplication problem, such as 2 x 24, we can think of this as finding 2 of a group of 24. Likewise, we are finding $\frac{2}{3}$ of a group of 24, so we can indicate this with a multiplication symbol. (In *Primary Mathematics* 5A, students will find a whole number times a fraction, and realize that 12 times groups of $\frac{2}{3}$ gives the same answer as $\frac{2}{3}$ of 12).

Activity

⇒ Josie filled 180 balloons with helium for a party. $\frac{5}{12}$ of them are red and $\frac{1}{2}$ of them are yellow. How many more yellow balloons are there than red balloons?

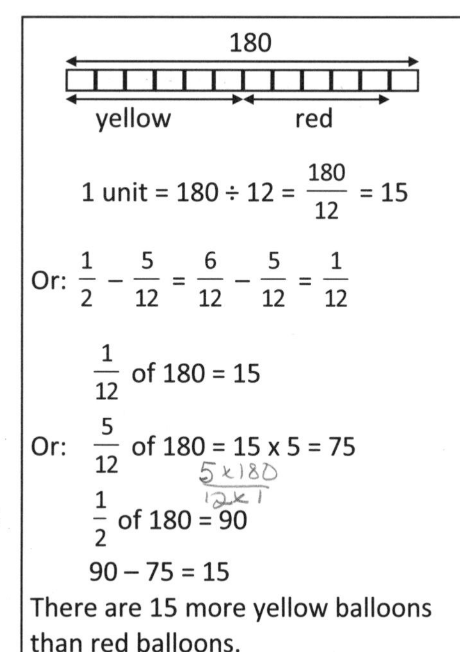

Discuss 3 methods. The first two shown at the right are similar. Ask your student which is easier. We might not even have thought of the first or second method without an understanding of bar models. Understanding how to use bar models is useful for making a problem easier to solve, even if we don't draw out the model. Encourage your student to use bar models as needed. They will make problems that she will encounter later in *Primary Mathematics* easier to solve. Note that we can find 180 ÷ 12 by simplifying the fraction $\frac{180}{12}$ or dividing 180 by 12.

Workbook

Exercise 16, pp. 100-101 (answers p. 119)

(5) Solve word problems 2

Discussion

Tasks 11-12, pp. 102-103

In these tasks, we are finding a different fraction than the one given in the problem. Three methods are shown to solve the problem in Task 11. Note that the calculations used for Method 2 and 3 are essentially the same. Method 3 allows students to visualize the problem. Ask your student which method is easiest. Allow him to use any method for solving word problems, but encourage the use of bar models.

11. $20 - 8 = \mathbf{12}$
    She had $**12** left
    $\frac{3}{5} \times 20 = 3 \times \frac{20}{5} = 3 \times 4 = \mathbf{12}$
    She had $**12** left.
    5 units = $20
    1 unit = $\$20 \div 5 = \$\frac{20}{5} = \mathbf{\$4}$
    3 units = 3 x $4 = $**12**
    She had $**12** left.
12. 8 units = 48
    1 unit = 48 ÷ 8 = **6**
    5 units = 6 x 5 = **30**
    There were **30** boys.

Activity

Have your student solve the following problems.

⇒ Sam had $126. He spent $\frac{4}{9}$ of it last week and $\frac{1}{3}$ of it this week. How much money does he have left?

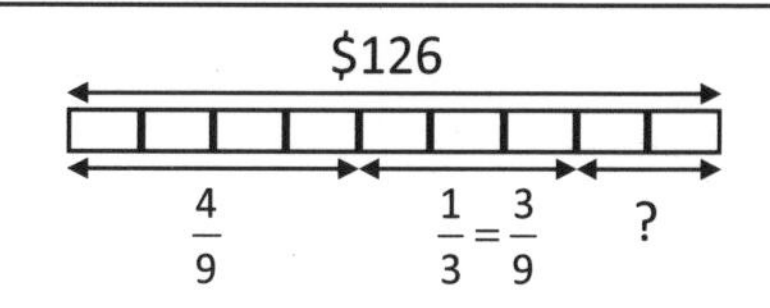

9 units = $126
1 unit = $126 ÷ 9 = $14
2 units = $14 x 2 = $28
He had $28 left.

⇒ Pam has $84. She spent $\frac{5}{7}$ of it on 5 identical shirts. How much do 3 such shirts cost?

$\frac{5}{7} \times \$84 = \$60$

5 shirts = $60
1 shirt = $60 ÷ 5 = $12
3 shirts = $12 x 3 = $36
3 shirts cost $36.

Workbook

Exercise 17, pp. 102-104 (answers p. 120)

Enrichment

⇒ A coat and a pair of pants together cost $91. The pants cost $\frac{2}{5}$ of what the coat cost. How much more did the coat cost than the pants?

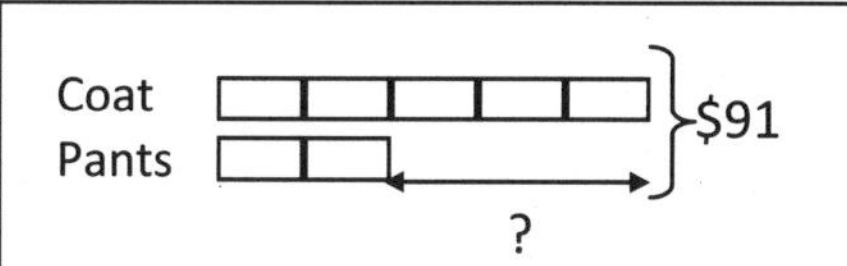

7 units = $91
1 unit = $91 ÷ 7 = $13
3 units = $13 x 3 = $39
The coat cost $39 more than the pants.

(6) Solve word problems 3

Discussion

Task 13, p. 104

In this task, we are not given the total amount. Instead, we are given the value of fraction, and asked to find the total amount. Make sure your student understands why the bar is divided into 5 parts, and why two parts are labeled with \$20. Ask her to also find how much money he had left.

13. 2 units = \$20
1 unit = \$20 ÷ 2 = \$**10**
5 units = \$10 x 5 = \$**50**
He had \$**50** at first.

Activity

⇒ Josie filled some balloons with helium for a party. $\frac{1}{5}$ of the balloons are red, $\frac{3}{10}$ of them are yellow, and the rest of them are blue. If 15 of the balloons are yellow, how many balloons did she fill with helium? How many of the balloons are blue?

To divide the bar into units, we need to first find equivalent fractions so all the parts are the same size. Since one fifth is the same as two tenths, we can use tenths.

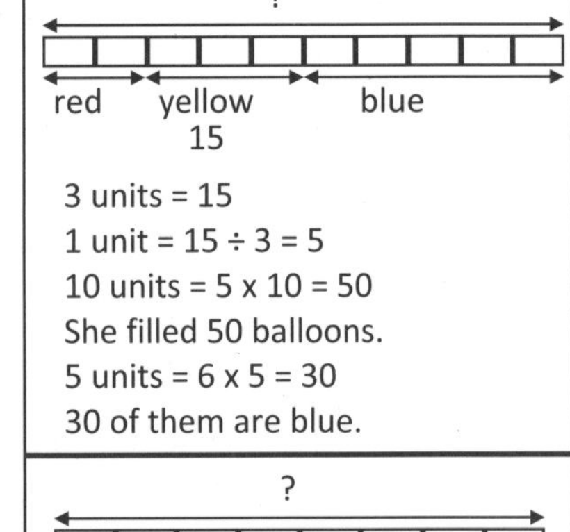

3 units = 15
1 unit = 15 ÷ 3 = 5
10 units = 5 x 10 = 50
She filled 50 balloons.
5 units = 6 x 5 = 30
30 of them are blue.

⇒ Sam read $\frac{5}{8}$ of a book on Monday and $\frac{1}{4}$ of it on Tuesday. If he read 48 more pages on Monday than on Tuesday, how many pages does he have left to read in the book? How many pages total is the book?

If we divide the bar into eighths, then he read 5 units on Monday and 2 on Tuesday. So he read 3 more units on Monday than on Tuesday, and has 1 more unit left to read.

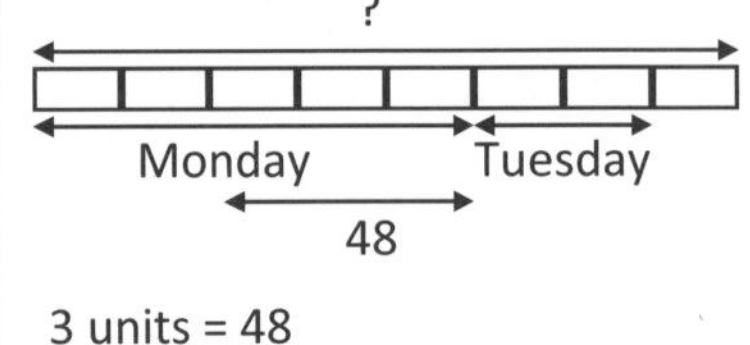

3 units = 48
1 unit = 48 ÷ 3 = 16
He has 16 more pages to read.
8 units = 16 x 8 = 128
The book is 128 pages long.

Workbook

Exercise 18, pp. 105-106 (answers p. 120)

Enrichment

⇒ Three sisters, Xenia, Yvette, and Zoe, share some pieces of candy. Yvette gets $\frac{1}{5}$ of the candies. Xenia gets 16 of the candies. Zoe gets 5 candies more than Yvette. How many pieces of candy did they share?

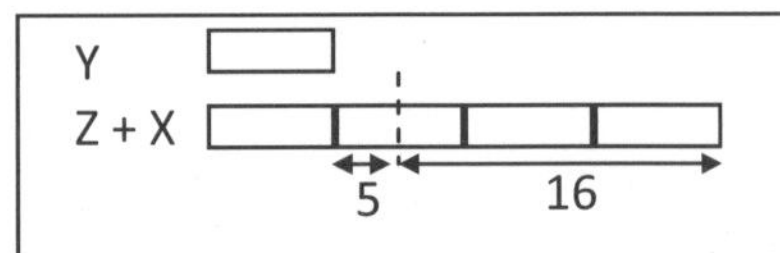

3 units = 5 + 16 = 21
1 unit = 21 ÷ 3 = 7
5 units = 7 x 5 = 35
They shared 35 pieces of candy.

⇒ Adam and Brendan want to share some pieces of candy equally. At first, Adam had $\frac{1}{6}$ of the candies and Brendan had 25 pieces of candy. How many pieces of candy did Brendan give Adam so they both had the same number of pieces of candy?

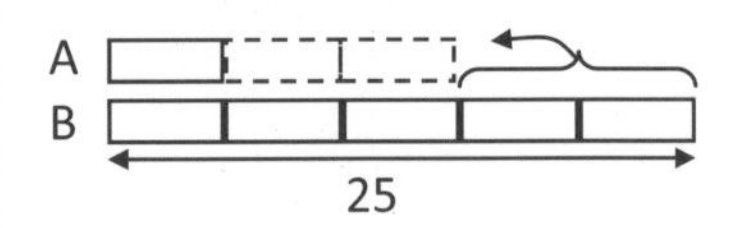

5 units = 25
1 unit = 25 ÷ 5 = 5
2 units = 5 x 2 = 10
Brendan gave Adam 10 candies.

(7) Solve word problems 4

Discussion

Task 14, p. 104

When we want to express one number as a fraction of a second number, the second number is the whole and the first number a part of the whole. In this problem, we are given the whole, and need to find the number of girls. Two methods are shown. Make sure your student understands both methods.

14. $\frac{15}{40} = \mathbf{\frac{3}{8}}$

$\mathbf{\frac{3}{8}}$ of the students in the class are girls.

$1 - \frac{5}{8} = \mathbf{\frac{3}{8}}$

$\mathbf{\frac{3}{8}}$ of the students in the class are girls.

Practice

Practice D, p. 105

2(c): Your student has not yet had problems involving fractions of a set where the result is not a whole number. You may need to remind him that $\frac{1}{8}$ of 180 = $\frac{180}{8}$. He can then simplify the improper fraction to a mixed number, or divide.

3: To solve this, you can tell your student to convert meters to centimeters and express the answer in compound units.

1. (a) 6 (b) 24 (c) 1 (d) 40
2. (a) 50 (b) 40 (c) $22\frac{1}{2}$ (d) 750
3. $\frac{3}{4}$ of 3 m = $\frac{3}{4}$ of 300 cm
   = 225 cm = **2 m 25 cm**
   The bookshelf was 2 m 25 cm long.
4. $\frac{3}{4}$ of 60 min = **45 min**
   She practices 45 min a day.
5. (a) $\mathbf{\frac{3}{5}}$ of the students do not wear glasses.
   (b) 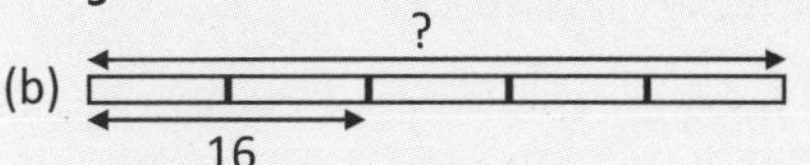

   2 units = 16
   1 unit = 16 ÷ 2 = 8
   5 units = 8 x 5 = **40**
   There are 40 students.
6. $\frac{1}{3}$ of 30 = **10**
   She had 10 eggs left.
7. 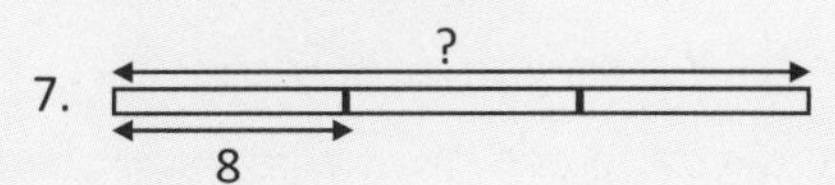

   1 unit = 8
   3 units = 8 x 3 = **24**
   She bought 24 picture cards.
8. 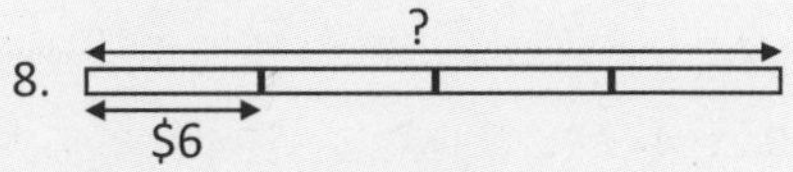

   4 units = $6 x 4 = **$24**
   He had $24 at first.

Workbook

Exercise 19, pp. 107-109 (answers p. 120)

Reinforcement

Extra Practice, Unit 3, Exercise 6, pp. 49-52

Tests

Tests, Unit 3, 6A and 6B, pp. 83-87

Review 3

Review

Review 3, pp. 106-109

Workbook

Review 3, pp. 110-116 (answers pp. 120-121)

Tests

Tests, Units 1-3 Cumulative Tests A and B, pp. 89-95

1. (a) 7003
   (b) 15,212
   (c) 62,409,200
   (d) –47
2. (a) forty-six thousand, six hundred
   (b) three hundred fifty-six thousand
   (c) four hundred seventy thousand, nineteen
   (d) five hundred two thousand, four hundred seventy-three
3. (a) 47,355; 74,355; 74;535; 75,435
   (b) 23,232; 23,322; 32,223; 33,222
   (c) –10, –5, 0, 2, 20, 30
4. (a) 16,060 (b) 69,516 (c) –101 (d) –99
5. –200 m
6. (a) 41,000 (b) 69,000 (c) 597,000
7. (a) 700,000 (b) 5,600,000 (c) 7,400,000
8. (a) 5385 (b) 2100 (c) 4590 (d) 456 R 2
9. (a) $f = 9$ (b) $f = 8$
   (c) $f = 3$ (d) $f = 2$
   (e) $f = 9$ (f) $f = 8$
   (g) $f = 3$ (h) $f = 2$
10. 11, 13, 17, 19
11. (a) $\underline{(4 + 32)} \div 6 + 15$
    $= \underline{36 \div 6} + 15$
    $= \underline{6 + 15}$
    $= 21$
    (b) $1000 - (\underline{20 \times 100} \div 8)$
    $= 1000 - (\underline{2000 \div 8})$
    $= \underline{1000 - 250}$
    $= 750$
12. (a) $\frac{3}{4}$ (b) $\frac{7}{9}$ (c) $\frac{11}{12}$
    (d) $\frac{1}{5}$ (e) $\frac{1}{4}$ (f) $\frac{1}{8}$
13. (a) $\frac{5}{8}$ (b) $\frac{1}{3}$ (c) $\frac{1}{6}$
    (d) $\frac{1}{2}$ (e) $\frac{3}{4}$ (f) $\frac{1}{2}$
14. (a) $\frac{4}{9}$ (b) $\frac{2}{3}$
15. $\frac{1}{12}, \frac{1}{3}, \frac{3}{5}, \frac{4}{4}, \frac{3}{2}$
16. (a) $\frac{\mathbf{6}}{10}$ (b) $\frac{3}{\mathbf{18}}$ (c) $\frac{\mathbf{2}}{3}$ (d) $\frac{2}{\mathbf{3}}$
17. (a) $\frac{4}{5}$ (b) $\frac{1}{6}$ (c) $1\frac{1}{3}$ (d) $2\frac{1}{4}$
18. (a) $3\frac{1}{3}$ (b) 3 (c) $4\frac{1}{2}$ (d) $3\frac{2}{7}$
19. (a) $\frac{11}{7}$ (b) $\frac{14}{5}$ (c) $\frac{25}{8}$ (d) $\frac{29}{10}$
20. $\frac{18\text{ m}}{4} = \mathbf{4\frac{1}{2}\text{ m}}$ Each piece is $4\frac{1}{2}$ m long.
21. 3 units = 150 g
    10 units = $\frac{150}{3}$ g x 10 = **500 g**
    She bought 500 g of oil.
22. 5 units = 20 ℓ
    6 units = $\frac{20}{5}$ ℓ x 6 = **24 ℓ**
    The tank's capacity is 24 ℓ.
23. $\frac{1}{4}$ x 100 = **25** There were 25 sandwiches left.
24. $\$\frac{20}{2} + \$\frac{24}{2} + \$\frac{36}{2}$ = \$10 + \$12 + \$18 = **\$40**
    (Or, add and then divide sum by 2.)
    She spent \$40.
25.

    (a) $\frac{1}{5}$ of the children in the choir are boys.
    (b) 1 unit = 8
    5 units = 8 x 5 = **40**
    There are 40 children in the choir.
    (c) There are 3 more units of girls than boys.
    3 units = 8 x 3 = **24**
    There are 24 more girls than boys in the choir.

Workbook

Exercise 12, pp. 90-92

1. Check answers.
2. (a) $\frac{2}{7}$ (b) $\frac{2}{3}$
   (c) $\frac{3}{4}$ (d) $\frac{3}{7}$
3. (a) $\frac{1}{2}$ (b) $\frac{5}{6}$
   (c) $\frac{1}{4}$ (d) $\frac{3}{8}$
   (e) $\frac{1}{5}$ (f) $\frac{2}{3}$
4. (a) $\frac{2}{5}$; $\frac{3}{5}$
   (b) $\frac{1}{2}$; $\frac{1}{6}$; $\frac{1}{3}$
   (c) $\frac{1}{2}$

Exercise 13, pp. 93-94

1. Check answers.
2. (a) 6 (b) 3
   (c) 8 (d) 6
3. (a) 5 (b) 5
   15 15
   (c) 7 (d) 3
   14 21
   (e) 4 (f) 4
   12 20

Exercise 14, pp. 95-97

1. (a) 2 (b) 3
   (c) 4 (d) 2; 4
   (e) 9 (f) 9
2. (a) 4 (b) 5
   (c) 5 (d) 3
   (e) 16 (f) 16
   (g) 15 (h) 15
3. (a) 10 (b) 15
   (c) 24 (d) 30
   (e) 32 (f) 45
   (g) 60 (h) 84

Exercise 15, pp. 98-99

1. (a) $\frac{20¢}{100¢} = \mathbf{\frac{1}{5}}$
   (b) $\frac{80\text{ cm}}{100\text{ cm}} = \mathbf{\frac{4}{5}}$
   (c) $\frac{25\text{ min}}{60\text{ min}} = \mathbf{\frac{5}{12}}$
2. (a) $\frac{8\text{ h}}{24\text{ h}} = \mathbf{\frac{1}{3}}$
   (b) $\frac{50}{90} = \mathbf{\frac{5}{9}}$
   (c) $\frac{45\text{ cm}}{100\text{ cm}} = \mathbf{\frac{9}{20}}$
3. $\frac{75\text{ m}}{100\text{ m}} = \mathbf{\frac{3}{4}}$
4. $\frac{16}{40} = \mathbf{\frac{2}{5}}$ $\frac{2}{5}$ of the children wear glasses.
5. $\frac{15}{40} = \mathbf{\frac{3}{8}}$ $\frac{3}{8}$ of the cars are battery operated.
6. $\frac{24}{60} = \mathbf{\frac{2}{5}}$ $\frac{2}{5}$ of the stamps are Canadian stamps.

Exercise 16, pp. 100-101

1. (a) $2 \times \frac{25}{5} = \mathbf{10}$ He gave his friends 10 cards.
   (b) $3 \times \frac{25}{5} = \mathbf{15}$ He had 15 cards left.
2. (a) $3 \times \frac{\$40}{8} = \mathbf{\$15}$ It cost $15.
   (b) $5 \times \frac{\$40}{8} = \mathbf{\$25}$ She had $25 left.
3. $\frac{36}{6} = \mathbf{6}$ 6 of the children wear glasses.
4. $2 \times \frac{24}{3} = \mathbf{16}$ She peeled 16 potatoes.
5. $3 \times \frac{\$48}{8} = \mathbf{\$18}$ She spent $18.

Workbook

Exercise 17, pp. 102-104

1. $4 \times \frac{\$25}{5} = \mathbf{\$20}$ She saved $20.
2. $3 \times \frac{45}{5} = \mathbf{27}$ He had 27 oranges left.
3. $\frac{\$48}{4} + \$14 = \mathbf{\$26}$ She spent $26.
4. $2 \times \frac{60}{5} - 15 = \mathbf{9}$ There are 9 girls.
5. $3 \times \frac{96}{4} = \mathbf{72}$ There were 72 males.
6. $5 \times \frac{144}{8} = \mathbf{90}$ 90 children are not running.

Exercise 18, pp. 105-106

1. (a) 1 unit = $\frac{\$42}{7} = \mathbf{\$6}$
   10 units = 10 x \$6 = **\$60**
   She had $60 at first.
   (b) 3 units = 3 x \$6 = **\$18**
   She saved $18.
2. (a) 7 units = $7 \times \frac{18}{3} = \mathbf{42}$
   There were 42 children.
   (b) 4 units = $4 \times \frac{18}{3} = \mathbf{24}$
   There were 24 girls.
3. 3 units = 3 x 6 kg = **18 kg**
   She bought 18 kg of flour.
4. 10 units = 10 x (\$9 ÷ 3) = **\$30**
   She had $30 at first.

Exercise 19, pp. 107-109

1. $\$25 - \$5 = \$20$; $\frac{\$20}{\$25} = \mathbf{\frac{4}{5}}$
   He had $\frac{4}{5}$ of his money left.
2. $100 - 60 = 40$; $\frac{40}{100} = \mathbf{\frac{2}{5}}$
   $\frac{2}{5}$ of the children were girls.
3. $6 \times 2\text{ m} = 12\text{ m}$; $\frac{12\text{ m}}{40\text{ m}} = \mathbf{\frac{3}{10}}$
   She used $\frac{3}{10}$ of the material.
4. (a) \$240 ÷ \$2 = **120**
   He sold 120 mangoes.
   (b) $160 - 120 = 40$; $\frac{40}{160} = \mathbf{\frac{1}{4}}$
   He had $\frac{1}{4}$ of his mangoes left.
5. 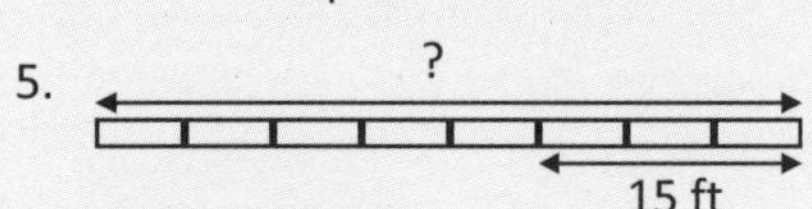

   3 units = 15 ft
   1 unit = 15 ft ÷ 3 = 5 ft
   8 units = 5 ft x 8 = **40 ft**
   She bought 40 ft of ribbon.
6. 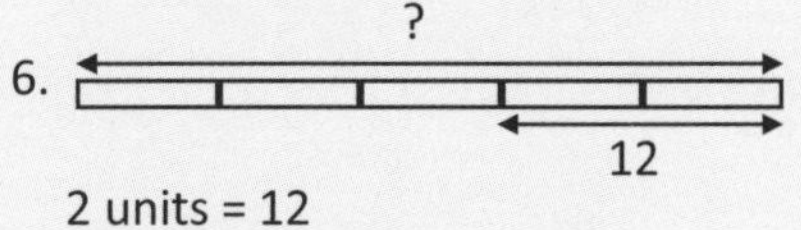

   2 units = 12
   3 units = (12 ÷ 2) x 3 = **18**
   He used 18 stamps.

Review 3, pp. 110-116

1. (a) sixty thousand, five hundred
   (b) forty-two million, eight hundred nineteen thousand
   (c) one hundred four
2. (a) 75,612
   (b) 80,002
3. 3000
4. 100,000
5. 80,036; 80,360; 83,060; 83,600; 86,300
6. 6300
7. 6, 12, 18, 24, 30
8. 4
9. (23 x 80) – 238 = 1840 – 238 = **1602**
10. (352 + 698) ÷ 5 = 1050 ÷ 5 = **210**
11. 622 R 2
12. –3; 2
13. $\frac{5}{12}, \frac{3}{4}, 1, \frac{7}{6}$

Workbook

14. $\frac{1}{4}$

15. $\frac{3}{4}$ of 80 = 3 x $\frac{80}{4}$ = 3 x 20 = **60**

16. 5345

17. $\frac{4 \text{ in.}}{12 \text{ in.}} = \mathbf{\frac{1}{3}}$ 4 in. is $\frac{1}{3}$ of a foot.

18. $\frac{7}{4}$

19. $4\frac{4}{5}$

20. 4 ÷ 5

21. (a) (11 x 4) ÷ (1 x 22) = 4 ÷ ____
    44 ÷ 22 = 4 ÷ ____
    2 = 4 ÷ **2**   (44 ÷ 22 = $\frac{44}{22}$)
    (b) 18 – (10 + 6) = ____ – 52
    2 = ____ – 52
    2 = **54** – 52   (2 + 52 = 54)
    (c) 20 x (18 ÷ 2) = 100 + ____
    180 = 100 + ____
    180 = 100 + **80**
    (d) (7 + 5) x (15 – 12) = ____ x 12
    12 x 3 = **3** x 12
    (e) (40 ÷ 8) + 3 = (4 x 2) – (1 x ____)
    5 + 3 = 8 – ____
    8 = 8 – **0**
    (f) 10 + (28 ÷ 7) = (8 x 5) – ____
    14 = 40 – ____
    14 = 40 – **26**   (40 – 14 = 26)
    (g) (72 ÷ 8) ÷ (3 + 6) = 3 ÷ ____
    9 ÷ 9 = 3 ÷ **3**

22. $1750 x 6 = **$10,500**
    She earns $10,500 in 6 months.

23. 1188 ÷ 6 = **198**
    She had 198 bags of muffins.

24. Nicole
    Tasha
    2000
    600
    Let Nicole's stickers be 1 unit.
    2 units = 2000 + 600 = 2600
    1 unit = 2600 ÷ 2 = **1300**
    Nicole has 1300 stickers.

25. Cost of chairs: $165 x 12 = $1980
    $2400 – $1980 = **$420**
    The table cost $420.

26. 3 x $\frac{60}{10}$ = **18**   There are 18 yellow roses.

27. 30 – 5 = 25; $\frac{25}{30} = \mathbf{\frac{5}{6}}$ She had $\frac{5}{6}$ of the eggs left.

28. $\frac{12}{3}$ = **4**   There were 4 sheets of paper left.

29. Adults
    Children
    2500
    5 units = 2500
    1 unit = 2500 ÷ 5 = 500
    4 units = 500 x 4 = 2000
    2000 – 1200 = **800**
    There were 800 women.

30. (a) $\frac{2}{5}$ lb + $\frac{3}{10}$ lb = $\frac{4}{10}$ lb + $\frac{3}{10}$ lb = $\mathbf{\frac{7}{10}}$ **lb**
    She used $\frac{7}{10}$ lb of the flour.
    (a) $\frac{4}{10}$ lb – $\frac{3}{10}$ lb = $\mathbf{\frac{1}{10}}$ **lb**
    She used $\frac{1}{10}$ lb more flour for the banana cake.

31. TV
    Stereo
    $1980
    $1200
    ?
    Total cost: (2 x $1980) + $1200 = $5160
    $5160 ÷ 4 = **$1290**
    Each person paid $1290.

32. Total grapefruits: 25 x 36 = 900
    900 – 28 –786 = **86**
    He had 86 grapefruits left.

33. Total orange juice:
    12 x 375 ml = 4500 ml = 4 ℓ 500 ml
    Amount left after filling the two 2-ℓ jugs:
    4 ℓ 500 ml – 4 ℓ = **500 ml**
    There was 500 ml of orange juice in the glass.

Unit 4 – Geometry

Chapter 1 – Right Angles

Chapter 2 – Measuring Angles

Objectives

- Relate quarter turn, half turn, three-quarter turn and whole turn to right angles.
- Classify angles as right, acute, or obtuse.
- Estimate and measure angles in degrees.
- Construct angles.

Vocabulary

- Angle
- Right angle
- Acute angle
- Obtuse angle
- Degrees
- Reflex angle
- Protractor

Material

- Folding meter stick or two strips of cardboard attached with a brad
- Set square (plastic triangle with a 90° angle)
- Protractor
- Compass

Notes

Students learned to identify angles and compare them to a right angle in *Primary Mathematics* 3B. In these two chapters your student will learn how to measure angles in degrees.

An **angle** is formed when two straight lines meet at a point. The point of intersection is the vertex of the angle, and the two lines are the sides of the angle.

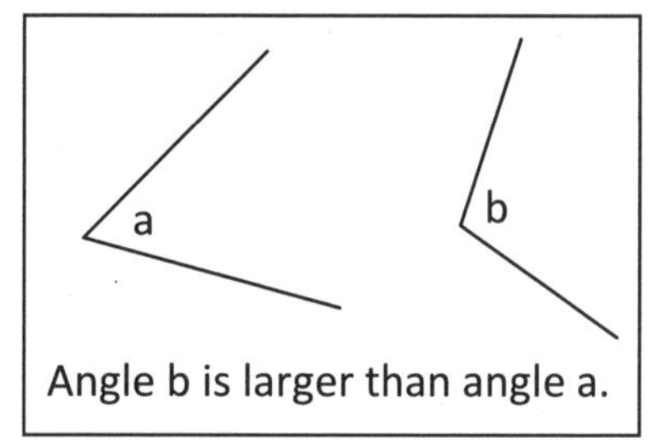

Angle b is larger than angle a.

The size of an angle is determined by how much either line is turned about the point where they meet. It does not depend on the length of the two sides.

The **degree** is derived from the Babylonian base 60 system. They may have assigned 360 degrees to a circle because they found that it took about 360 days for the sun to complete one year's circuit across the sky. 360 is conveniently divisible by 2, 3, 4, 5, 6, 8, 9, 10, 12, 15, 18, and 20, so the degree is a nice unit to use to divide the circle into an equal number of parts. 180 degrees is half of the way around a circle, 120 degrees is one third of a the way around a circle, 90 degrees is one fourth of the way around a circle, and so on.

The abbreviation for degree is a superscript O (e.g., 90°). Angles are measured with a **protractor**. A 90° angle is called a **right angle**. An angle smaller than 90° is called an **acute angle**, and one larger than 90° but less than 180° is called an **obtuse angle**. An angle larger than 180° is called a **reflex angle**. Do not let vocabulary terms impede mathematical understanding.

(1) Identify angles

Discussion

Concept p. 110

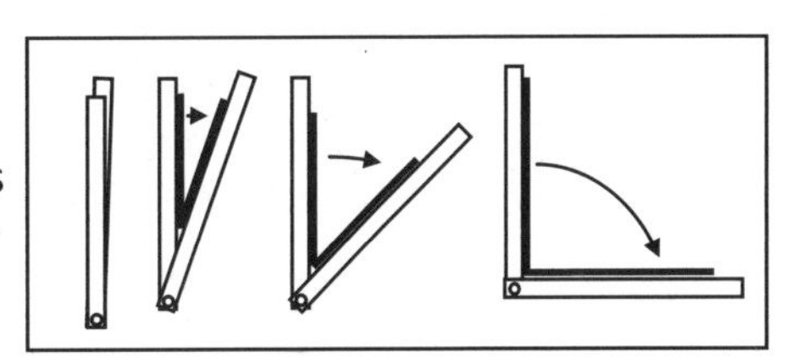

Use two strips of cardboard fastened with a brad at the bottom or a folding meter stick to illustrate the concepts pictured here concretely so that your student can better visualize that an angle is measured by the degree of turning.

Remind your student that an angle is formed when two straight lines meet at a point. The larger the angle, the more one side has to be turned away from the other. The size of the angle depends on the degree of turning. A complete turn is a circle. One fourth of a turn is a *right angle*.

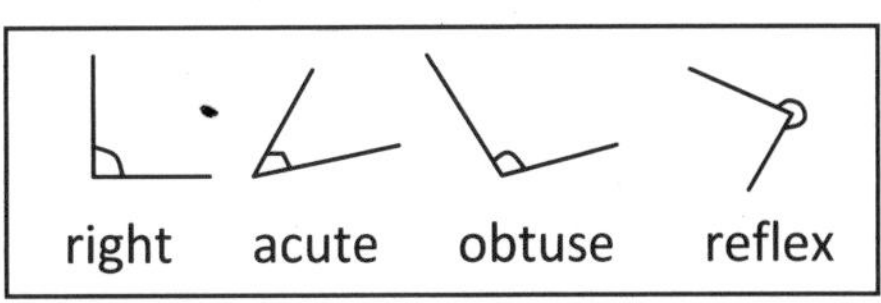

Tell your student that an angle smaller than a right angle is called an *acute angle*, an angle larger than one right angle but smaller than two right angles is called an *obtuse angle*, and an angle larger than two right angles is called a *reflex angle*. Point out that every angle actually has an opposite angle, and a reflex angle is opposite the other three angles. Draw some angles and ask him to identify which type of angle each one is.

Tasks 1-2, p. 111

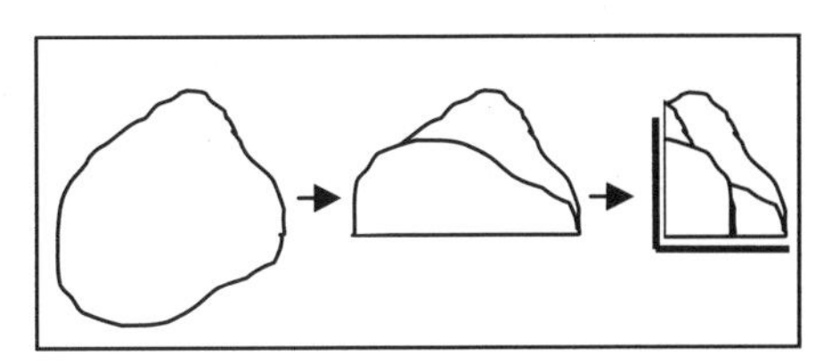

Folding a piece of paper is redundant, since each corner is already a right angle. You can instead have your student tear a rough circle out of paper and fold that. This emphasizes that a right angle is a fourth of a turn.

Activity

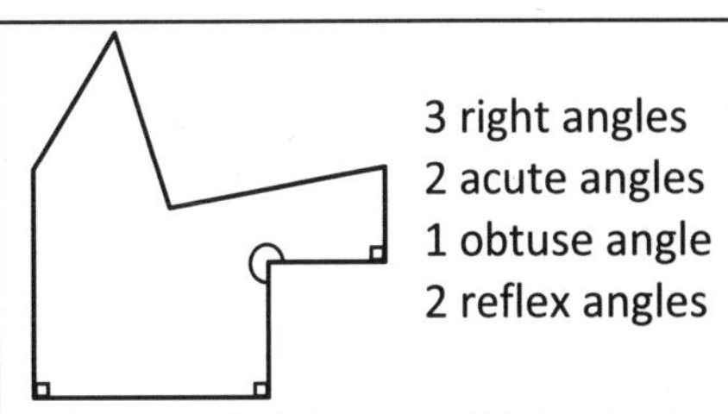

Draw a figure similar to the one shown here and ask your student how many angles are acute angles, right angles, obtuse angles, or reflex angles. Remind her that usually when we are asked to look at the angles of a shape like this, they are supposed to just look at the inside angles.

Practice

Task 3, p. 111

> 3. Angles A, B, and C are right angles.
> Angle E is an acute angle.
> Angles D and F are obtuse angles.

Workbook

Exercise 1, pp. 117-120 (answers p. 132)

Reinforcement

Extra Practice, Unit 4, Exercise 1, pp. 61-62

Test

Tests, Unit 4, 1A and 1B, pp. 97-101

(2) Measure angles smaller than 180°

Activity

Show your student a protractor and discuss the markings on it. Most protractors show measurements from 0° to 180° around the edge of the half circle, going clockwise, with one mark for each degree. They then have the degrees going counter-clockwise under that, with one mark for every ten degrees. Since the protractor is transparent, it can be flipped over to measure in either direction for either set of marks (outer or inner circles). Draw a large angle and show your student how to measure it, lining up the vertex and one side correctly (not at the bottom edge of the protractor, but at the line slightly up from the base). Help him to read the measurement. Have him measure the same angle, this time lining the baseline of the protractor up with the other side of the angle. Draw various angles and ask him to measure them.

Discussion

Concept p. 112

In these drawings, the angle is being measured using the inside circle on the protractor. You may want to point out that if we subtract the measurement shown on the outside circle from 180, we get the same measure as the value on the inside circle. If we want to measure an angle that is not close to a multiple of ten, we would use the marks on the outside circle, and starting from 0 by flipping the protractor or by lining up the other side of the angle with the baseline on the protractor.

Draw your student's attention to the abbreviation for angle. Tell her that sometimes, rather than labeling the angle, we label a point on each side and a point at the vertex with letters. Then we can name the angle by the letters for the points. Show her an example, such as the one at the right. Tell her that usually, when an angle is named, it refers to the angle that is less than 180°, not the opposite reflex angle. Also, a specific point is not always drawn, just the letter drawn near the line.

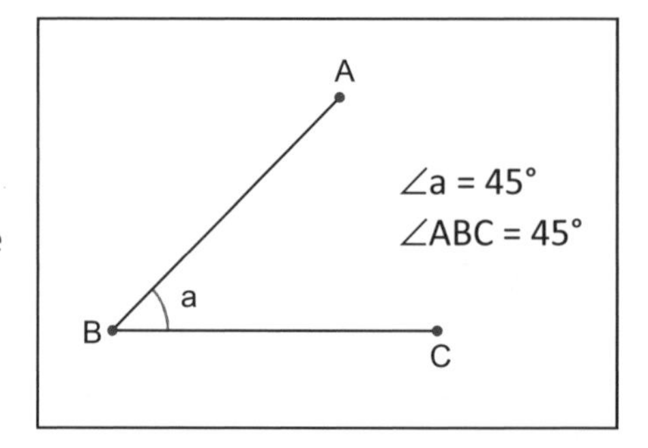

Practice

Task 1, p. 113

1. $\angle p = \mathbf{20°}$ $\angle q = \mathbf{140°}$

Workbook

Exercise 2, pp. 121-122 (answers p. 132)

It is difficult to measure angles precisely with a simple protractor. Your student's answers may vary from the answer key by several degrees. Accept answers within 3-5 degrees of those in the answer key.

(2) Draw angles smaller than 180°

Activity

Show your student how to draw angles. First, draw a line and mark a point on it, which can be at one end, where the vertex or point of the angle will be. Place the center point of the protractor on that point and line up the base line of the protractor directly on the top of the drawn line. Mark a point on the curved edge of the protractor corresponding to the required degrees. Make sure that the degrees start from 0; if necessary, flip the protractor. Remove the protractor, and then use the straight edge to connect the vertex with the marked point. Have him draw a right angle and several angles larger or smaller than a right angle.

Point out the 45° on the protractor and let your student see that it is half of the way to a right angle (or an eighth of the way around a full circle). 30° and 60° are a third and two thirds of the way to a right angle respectively. Similarly, 120°, 135°, and 150° are a third of the way, half the way, and two thirds of the way from a right angle to a half turn, or 180°. Have her draw a 30°, 45°, and 60° angle and compare their relative sizes. Then draw some angles and have her use these measures as a benchmark to first estimate the angles before measuring them.

Draw some angles with sides that are too short to accurately measure with a protractor. Tell your student that in order to measure them accurately with the protractor, he will have to extend the sides. Guide him in extending them correctly, using the bottom edge of the protractor as a ruler.

Practice

Tasks 2-3, p. 113

2. $\angle ABC = \mathbf{77°}$ $\angle PQR = \mathbf{118°}$

3. Check angles

Workbook

Exercise 3, pp. 123-127 (answers p. 132)

Enrichment

Draw a large rectangle and ask your student what type of angle is formed in each corner and what each would measure. All corners of a rectangle are right angles and so are 90°.

Draw a diagonal, a line from one corner to the opposite corner. Ask your student to measure both of the angles formed in one corner, and then add them together. The sum should be 90°.

Tell your student that since the corner angle is 90°, the sum of any angles formed by lines coming out from the corner has to be 90°.

Draw some right angles, divide the angle with a third line, label one angle with an approximate measurement in degrees, and have your student find the unknown angle without measuring.

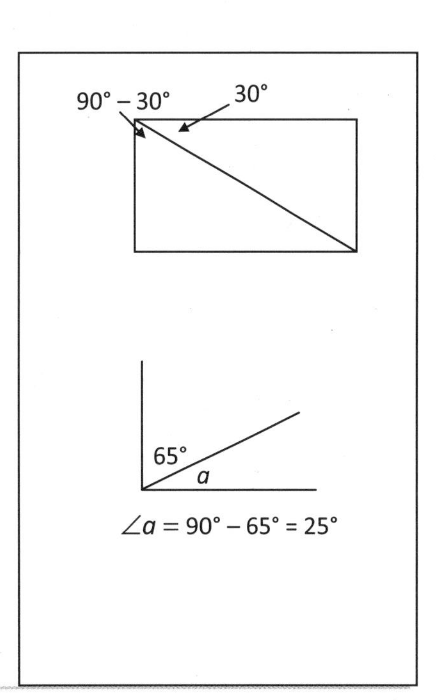

(4) Measure and draw angles greater than 180°

Discussion

Task 4, p. 114

This task relates the turnings in the previous chapter to actual degrees. Point out a quarter turn is equal to a right angle, which is 90°, so a half turn which is two right angles is 2 x 90°= 180°, a three-quarter turn which is three right angles is 3 x 90°= 270°, and a full turn which is four right angles is 4 x 90°= 360°.

Tasks 5-6, pp. 114-115

5. $\angle x =$ **240°**

6. $\angle y =$ **320°**

Since a protractor only measures up to a half-circle, we need a way to measure angles greater than 180°. These two tasks show two methods for measuring such angles. Before going through these tasks, you can draw an angle and ask your student to measure the reflex side. See what method she comes up with. Then discuss the tasks as reinforcement.

Method 1: Since a full circle is 360°, the reflex angle and the opposite angle together add up to 360°. So all we need to do is measure the smaller angle, and subtract it from 360°.

$\angle r = 360° - \angle a$

Method 2: Since a protractor measures a half circle, we can imagine or draw a straight line out from one of the sides. The reflex angle is the rest of the turn from the smaller angle to 180° and then another 180°.

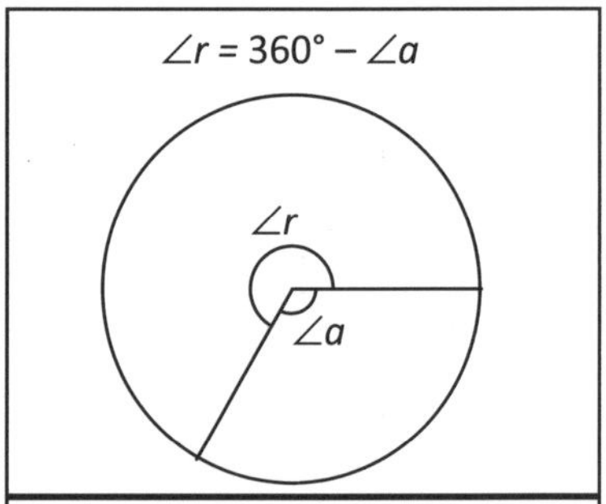

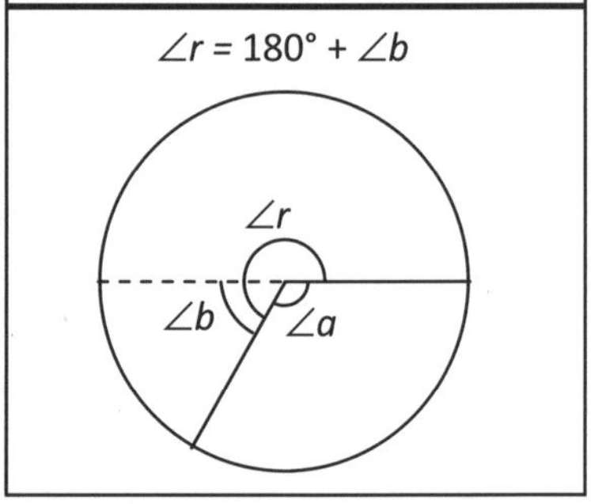

$\angle r = 180° + \angle b$

Task 7, p. 115

The strategies used to measure reflex angles can also be used to draw them. Task 7 shows a method that corresponds to Method 2 above. Subtract the desired angle from 360° to draw the angle opposite the reflex angle.

Practice

Task 8, p. 115

8. $\angle a =$ **250°** $\angle b =$ **240°** $\angle b =$ **290°**

It may help to trace these angles so your student can extend the sides in order to measure them.

Workbook

Exercise 4, pp. 128-131 (answers p. 132)

Reinforcement

Extra Practice, Unit 4, Exercise 2, pp. 63-68

Test

Tests, Unit 4, 2A and 2B, pp. 103-109

Chapter 3 – Perpendicular Lines

Chapter 4 – Parallel Lines

Objectives

- Identify perpendicular lines.
- Construct perpendicular lines.
- Identify parallel lines.
- Construct parallel lines.

Vocabulary

- Perpendicular lines
- Parallel lines
- Intersecting lines
- Horizontal line
- Vertical line

Material

- Set square (plastic triangle with a 90° angle)
- Ruler
- Square grid paper (appendix p. a17)
- Appendix pp. a22-24

Notes

In *Primary Mathematics* 3B students were briefly introduced to parallel lines and used them to identify parallelograms. In these two chapters, your student will learn about both perpendicular and parallel lines and how to construct them with a set square and ruler.

Intersecting lines are lines that cross each other. The point where they cross is called the point of intersection. **Perpendicular lines** intersect at right angles to each other. **Parallel lines** are lines that will never cross each other no matter how far they are extended. The distance between all points on the two lines is always the same. It is not necessary at this point to require a formal definition of lines and line segments, but since a line theoretically goes on indefinitely in either direction, your student will actually be dealing with line segments in these chapters.

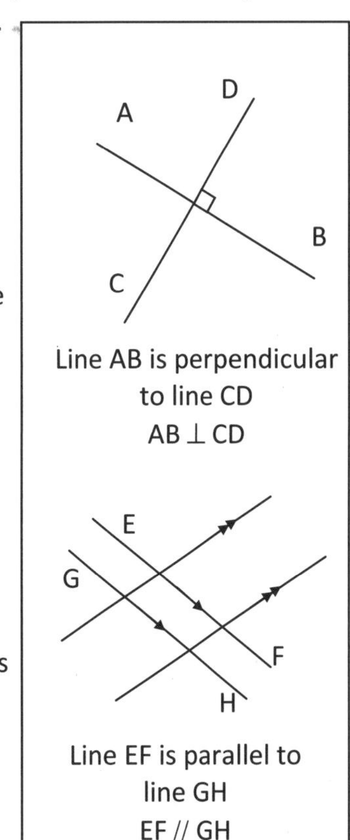

Line AB is perpendicular to line CD
AB ⊥ CD

Line EF is parallel to line GH
EF // GH

In drawings, perpendicular lines are indicated by a little square at the intersection, and parallel lines are indicated by little arrowheads. If the drawing has more than one set of parallel lines, lines parallel to each other are indicated by the same number of arrowheads.

Two points on a line labeled with capital letters, such as A and B, are used to name the line. Often, the exact point is not specified. The symbol ⊥ means perpendicular and // means parallel. So for the diagrams at the right, we can write AB ⊥ CD and EF // GH.

A **horizontal line** is a line that is parallel with the horizon. On paper, we take the bottom or top edge to be the horizon, so a horizontal line is parallel with the bottom edge of a page. A **vertical line** is a line that is perpendicular to a horizontal line. It is parallel with the sides of the paper.

(1) Identify perpendicular lines

Discussion

Concept pp. 116-117

Page 116 introduces perpendicular lines concretely by using pictures of objects. Have your student look at the thick dark lines in the pictures and tell you what angle the lines make with each other, marked with a little square. Tell him that lines that intersect with a right angle are called *perpendicular lines*. You can ask him to find other perpendicular lines in the environment.

For the pictorial representations at the top of p. 117, point out that all four angles formed by the intersection of the two perpendicular lines are right angles. Also, a line is perpendicular to another line even if it does not cross the other line, but would form a right angle if either of the lines were extended. For example, all the lines in the brick wall on p. 116 going up and down are perpendicular to the heavy dark line going from left to right.

Point out the abbreviation for perpendicular lines.

Task 1, p. 117

Have your student write down the perpendicular lines using the correct symbols.

1. (a) CD ⊥ BC; AD ⊥ AB
   (b) EJ ⊥ HJ; EF ⊥ EJ; GH ⊥ FG

Practice

Ask your student to identify perpendicular lines on copies of appendix pages 22 and 23 by drawing a little square in the intersections.

Workbook

Exercise 5, pp. 132-133 (answers p. 132)

(2) Draw perpendicular lines

Discussion

Task 2 p. 118

This task shows how to construct a line perpendicular to a given line through a given point using a set-square. Show your student the steps with an actual set-square. Line up one side of the set-square with the line, slide it along until the perpendicular side of the set-square touches the point, and then draw a line from the given line through the point along the side of the set-square. You can put the tip of the pencil on the point and slide the set-square along until it rests against the pencil.

Draw some lines and points and have your student construct a perpendicular line through the points.

Task 3, p. 118

This task shows how perpendicular lines can be drawn using square graph paper as a guide, since the lines on the graph paper intersect at right angles. All the lines in the diagram are perpendicular. Discuss how we can determine if they are perpendicular without using a set square.

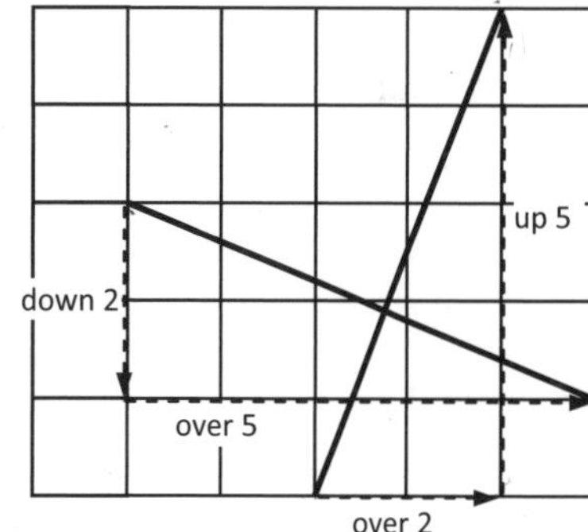

We could count the steps from where one line intersects a corner to where it intersects another corner. If one line goes over and down the same amount as the other line goes up and over, then they are perpendicular. For example, in the drawing at the right one line goes to the right 5 squares and down 2 squares from one grid corner to the next, and the other goes up 5 squares and to the right 2 squares.

Have your student draw some perpendicular lines on graph paper (such as appendix p. a17).

Workbook

Exercise 6, pp. 134-135 (answers p. 132)

Reinforcement

Extra Practice, Unit 4, Exercise 3, pp. 69-70

Test

Tests, Unit 4, 3A and 3B, pp. 111-116

(3) Identify parallel lines

Discussion

Concept pp. 119-120

Page 119 introduces parallel lines concretely by using pictures of objects. Have your student look at the thick dark lines in the pictures that are marked with arrows. These lines are always the same distance from each other. So if they were extended in either direction they would never cross each other. Such lines are called *parallel lines*. Ask her for other examples of parallel lines in the environment.

For the drawings at the top of p. 120, point out that if the pair of lines on the right were extended to the left, they would eventually cross each other. They are therefore not parallel lines.

Point out the abbreviation for parallel lines.

You can discuss the terms horizontal and vertical. *Vertical lines* are parallel to each other, as are *horizontal lines*. Vertical lines are perpendicular to horizontal lines.

Task 1, p. 120

Have your student write down the perpendicular and parallel lines using the correct symbols.

1. ST ⊥ RS
   PT // QR

Practice

Ask your student to identify parallel lines on copies of appendix page a24. She can draw one arrow on all the lines that are parallel to each other, two arrows on lines that are parallel to each other but not to the first set, and so on.

Workbook

Exercise 7, pp. 136-137 (answers p. 132)

(4) Draw parallel lines

Discussion

Task 2 p. 121

This task shows how to construct a line parallel to a given line through a given point using a set -square and a ruler. Show your student the steps with a set-square and a ruler. Line up one side of the right angle of the set-square with the line, line up the ruler with the other side, slide the set-square along the ruler until the first side intersects with the point, and then draw a line along the first side.

Have your student practice drawing lines parallel to a given line through a point. He can also draw squares and parallelograms.

Task 3, p. 121

This task shows how parallel lines can be drawn using square graph paper as a guide, since the vertical or horizontal lines on the graph paper are parallel to each other. All the lines in this task are parallel. Discuss how the squares in the background let us verify that the lines are indeed parallel.

For lines that coincide with the sides of the squares it is easy to see that they are parallel. For lines that slant, lines parallel to each other slant the same amount. That is, if we count over and up from where one line intersects a corner to the next place it intersects a corner, the number of squares over and up are the same for both parallel lines.

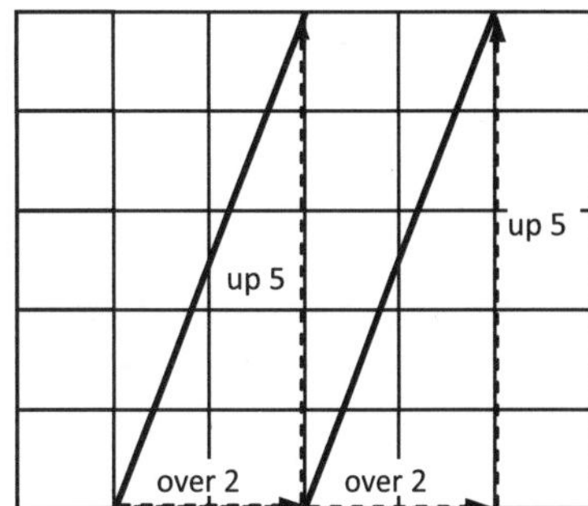

Have your student draw some perpendicular lines on graph paper (such as appendix p. a17).

Workbook

Exercise 8, pp. 138-139 (answers p. 132)

Reinforcement

Extra Practice, Unit 4, Exercise 4, pp. 71-72

Test

Tests, Unit 4, 4A and 4B, pp. 117-124

Workbook

Exercise 1, pp. 117-120

1. A: 3 B: 4
   C: 4 D: 5
   E: 6 F: 6

2. A: 2 B: 1
   C: 3 D: 3
   E: 3 F: 3

3.

Right angles	∠c, ∠d
Acute angles	∠a, ∠e
Obtuse angles	∠b, ∠f

4.

Right angles	∠a, ∠b, ∠c, ∠d, ∠h, ∠l, ∠m, ∠n, ∠p, ∠y
Acute angles	∠e, ∠j, ∠s, ∠q, ∠u, ∠w
Obtuse angles	∠f, ∠g, ∠i, ∠k, ∠o, ∠r, ∠t, ∠v, ∠x

Exercise 2, pp. 121-122

1. ∠b = 70° ∠c = 50°
   ∠d = 30° ∠e = 88°

2. ∠a = 100° ∠b = 120°
   ∠c = 140° ∠d = 160°
   ∠e = 110° ∠f = 130°

Exercise 3, pp. 123-127

1. ∠a = 62° ∠b = 90°
   ∠c = 105° ∠d = 65°
   ∠e = 63° ∠f = 122°

2. ∠a = 82° ∠b = 123°
   ∠c = 119° ∠d = 67°
   ∠e = 140° ∠f = 142°

3-6. Check angles

Exercise 4, pp. 128-131

1. ∠b = 360° – **160°** = **200°** ∠c = 360° – **130°** = **230°**
   ∠d = 360° – **113°** = **247°** ∠e = 360° – **90°** = **270°**

2. ∠a = 360° – **60°** = **300°** ∠b = 360° – **25°** = **335°**
   ∠c = 360° – **65°** = **295°** ∠d = 360° – **37°** = **323°**
   ∠e = 360° – **32°** = **328°** ∠f = 360° – **79°** = **281°**

3.

Angle	a	b	c	d	e	f
Measure	60°	213°	250°	29°	97°	294°

4-5. Check angles

Exercise 5, pp. 132-133

1. not perpendicular
   not perpendicular
   perpendicular
   perpendicular
   not perpendicular

2. XY ⊥ **XZ**
   PR ⊥ **QR**
   HK ⊥ JK, JK ⊥ IJ
   AE ⊥ DE, AB ⊥ BC, BC ⊥ CD

Exercise 6, pp 134-135

Check drawings.

Exercise 7, pp. 136-137

1. AB // **EF**
   MN // YZ
   PS // QR
   KN // LM, MN // KL

2.

Parallel lines	Perpendicular lines
AB // CD, EF // GH, WZ // XY, NO // LM, LO // MN	SR ⊥ PQ, IJ ⊥ JK, XW ⊥ XY, WZ ⊥ XW

Exercise 8, pp. 138-139

Check drawings.

Chapter 5 – Quadrilaterals

Objectives

- Identify common polygons (3, 4, 5, 6, and 8 sides).
- Identify parallelograms and trapezoids.
- Recognize rectangles, squares, and rhombuses as types of parallelograms.
- Identify various attributes of different types of quadrilaterals.
- Find the unknown lengths of sides of parallelograms given the lengths of other sides.

Vocabulary

- Polygon
- Triangle
- Quadrilateral
- Pentagon
- Hexagon
- Octagon
- Parallelogram
- Rhombus
- Trapezoid

Notes

In *Primary Mathematics* 3B, students learned to identify and name polygons with 3, 4, 5, 6, and 8 sides, and identify different kinds of parallelograms, including squares, rectangles, and rhombuses. In this chapter your student will review the names of common polygons and the properties of parallelograms. He will learn to identify trapezoids and other quadrilaterals by determining which sides are parallel and which are equal in length. In *Primary Mathematics* 5, students will learn the angle properties of these quadrilaterals.

Angles	Name
3	Triangle or Trigon
4	Quadrilateral or Tetragon
5	Pentagon
6	Hexagon
7	Heptagon
8	Octagon
9	Nonagon or Enneagon
10	Decagon
11	Hendecagon or Undecagon
12	Dodecagon
20	Icosagon
100	Hectogon
10000	Myriagon

A **polygon** is a closed figure formed by straight sides. In a closed figure, there is no way to get from the inside of the figure to the outside without crossing a line. Figure A at the right is a polygon, but figures B and C are not.

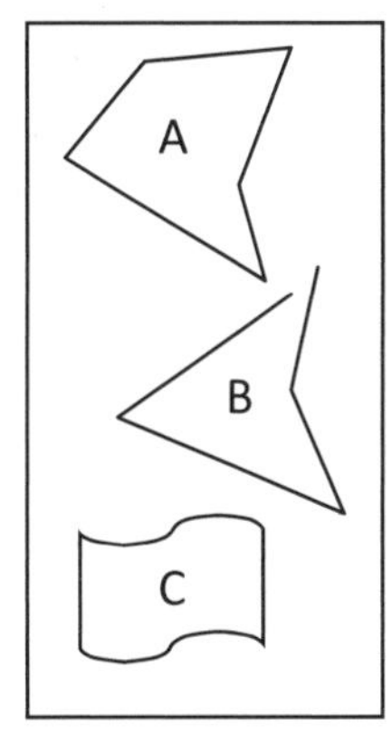

Poly- means many, and -gon means angle, so a polygon is named for the number of angles (which is the same as the number of sides). A pentagon has 5 sides, a hexagon has 6 sides, and an octagon has 8 sides. A triangle could also be called a trigon, and a quadrilateral a tetragon, but those names are not commonly used. Names of other polygons are given at the right. Your student is not required to know the names of polygons with 7 or more than 8 sides.

A **quadrilateral** is a 4-sided closed figure.

A **parallelogram** is a quadrilateral with both pairs of opposite sides parallel. The opposite sides are equal in length.

A **trapezoid** is a quadrilateral with only one pair of parallel sides.

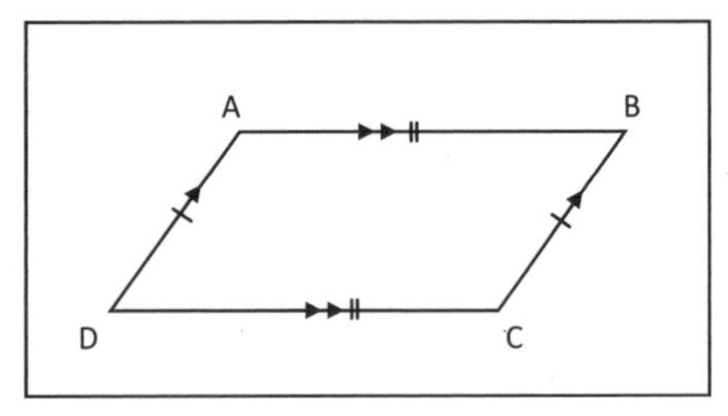

Primary Mathematics follows the common definition of a trapezoid: a quadrilateral with only two sides parallel. However, the restriction that a trapezoid has only one pair of parallel sides is not yet universally used. Some texts will say a trapezoid has "at least" one pair of parallel sides, and so might refer to a parallelogram (which has two pairs of parallel lines) as a trapezoid.

In drawings, arrows are used to indicate which sides are parallel and small cross marks are used to indicate which sides are equal in length. If the drawing has more than one pair of parallel sides or more than one set of equal sides, then the sides with the same number of marks are parallel or equal to each other. In the diagram of the parallelogram at the right, the markings indicate that AB = CD, DA = BC, AB // CD, and AD // BC.

A **rhombus** is a parallelogram with four equal sides. A square is a special kind of rhombus where all the angles are right angles. It is also a special kind of rectangle where all the sides are equal. A rectangle is a special type of parallelogram where all the angles are right angles.

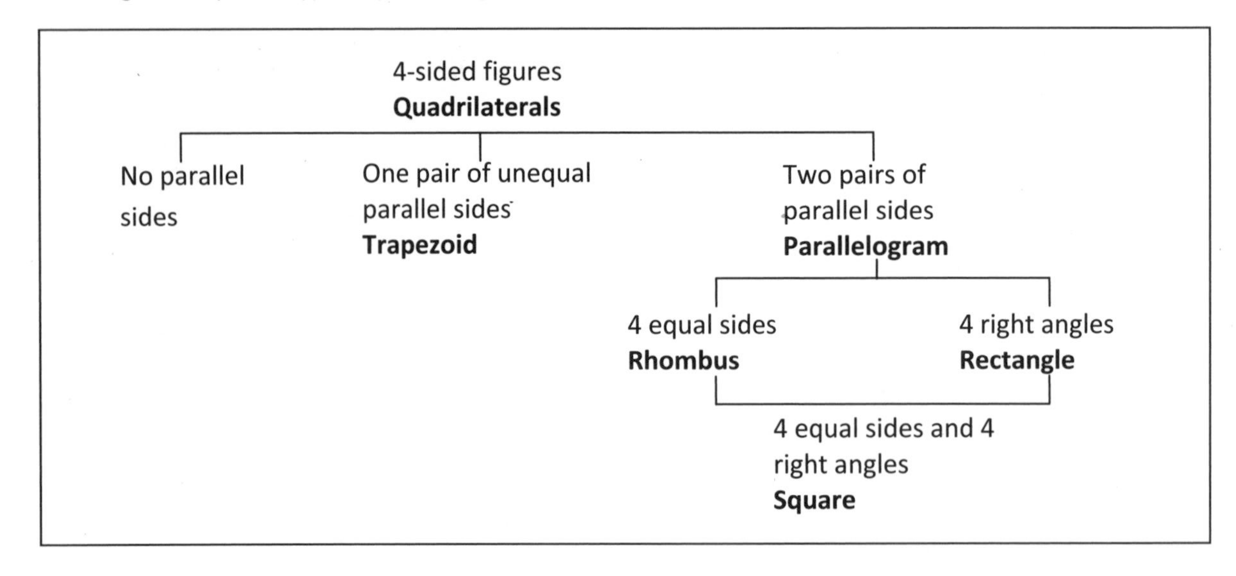

(1) Identify quadrilaterals

Discussion

Concept p. 122

This page provides a review of polygons and quadrilaterals. Tell your student that in a closed figure, there is no way to get from the inside of the figure to the outside without crossing a line. You can ask her to draw some examples of open figures or closed figures that are not polygons.

As you go through the questions on this page, make sure your student understands that the examples of the 4-sided figures include two basic types of quadrilaterals, ones with two pairs of parallel sides (parallelograms) and ones with one pair of parallel sides (trapezoids). This is the first time he has been given the name for trapezoids in this curriculum, so you will have to supply that name. A third type of quadrilateral has no parallel sides. You can ask him to draw a quadrilateral that is neither a parallelogram nor a trapezoid.

A triangle has **3** sides.
A quadrilateral has **4** sides.
A pentagon has **5** sides.
A hexagon has **6** sides.
An octagon has **8** sides.
B, C, D, E, F, and G are quadrilaterals.
F and G have only one pair of parallel sides.
They are called **trapezoids**.
B, C, D, and E have two pairs of parallel sides.
They are called **parallelograms**.
D and E have 4 equal sides.
They are called **rhombuses**.
C and D have 4 right angles.
They are called **rectangles**. (D is also called a square).

Tasks 1-5, pp. 123-124

These tasks review the properties of some quadrilaterals.

1: Refer back to the examples on the previous page to compare the 3 trapezoids. The "slant" of the two non-parallel sides can vary.

2: Point out the markings indicating which pairs of lines are equal and parallel.

1. AB // CD
   $\angle$BAC and $\angle$CDB are obtuse.
   $\angle$ABD and $\angle$ACD are acute.
2. EF // GH, EH // FG
   EF = GH, EH = FG
3. JK = 15 cm
   KL = 15 cm
   LM = 15 cm
4. WXYZ also called a **rectangle**.
   WX $\perp$ XY, XY $\perp$ YZ, YZ $\perp$ ZW, ZW $\perp$ WX
   YZ = 10 in.
   ZW = 3 in.
5. QRST is not a trapezoid because it has more than one pair of parallel sides.
   QRST is a parallelogram because it has 2 pairs of parallel sides.
   QRST is a rectangle because it has 4 right angles.
   QRST is a rhombus because it is a parallelogram with equal sides.
   QRST is a **square**.

Practice

Task 6, p. 124

6. ZY = 8 in.   DC = 7 cm
   WZ = 3 in.
   TU = 2 ft   SR = 5 cm

Workbook

Exercise 9, pp. 140-141 (answers p. 145)

Reinforcement

Extra Practice, Unit 4, Exercise 5, pp. 73-74

Test

Tests, Unit 4, 5A and 5B, pp. 125-131

Chapter 6 – Triangles

Objectives

- Identify equilateral, isosceles, and scalene triangles.
- Identify right, acute, and obtuse triangles.
- Find the unknown lengths of sides of triangles given the lengths of other sides.

Vocabulary

- Isosceles
- Equilateral
- Scalene
- Right triangle
- Acute triangle
- Obtuse triangle

Notes

In *Primary Mathematics* 3B, students learned to identify and name three kinds of triangles based on the length of their sides. In this chapter, your student will learn to classify triangles based on their angles.

Triangles can be classified in two ways.

1. According to angles:

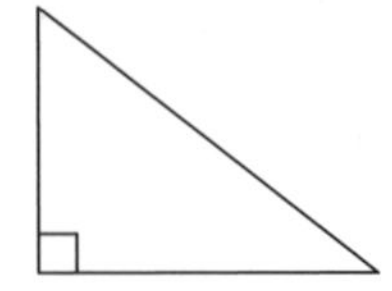

A **right triangle** has a right angle (90°).

An **acute triangle** has 3 acute angles (smaller than 90°).

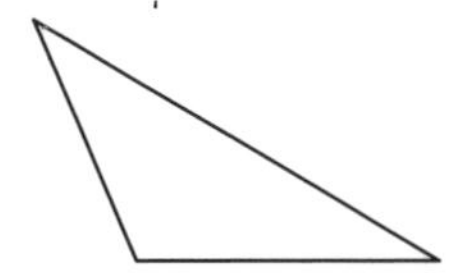

An **obtuse triangle** has one obtuse angle (greater than 90°).

2. According to sides:

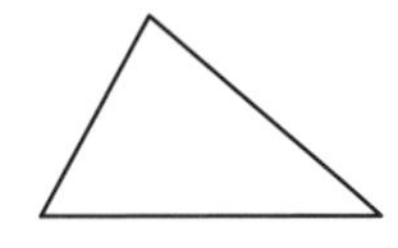

A **scalene triangle** has all sides different in length.

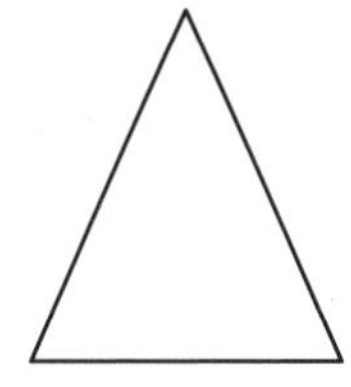

An **isosceles triangle** has two equal sides.

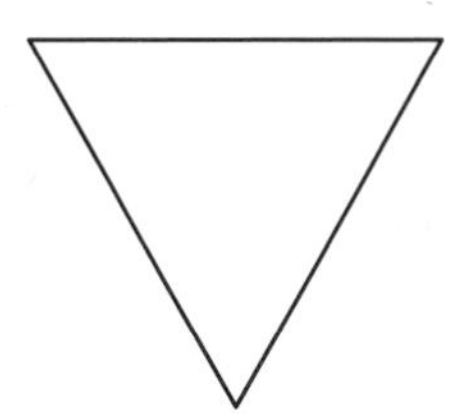

An **equilateral triangle** has three equal sides.

In *Primary Mathematics*, an equilateral triangle is considered to also be an isosceles triangle.

(1) Identify triangles

Discussion

Concept p. 125-126

These pages provide a review of *equilateral*, *isosceles*, and *scalene* triangles.

Remind your student that an equilateral triangle is also an isosceles triangle, since it also has 2 equal sides. So an equilateral triangle is a special type of isosceles triangle where the third side is also equal to the other two. However, he should call it by the more specific name of equilateral triangle.

You may want to discuss the origin of the names for triangles and the meanings of the prefixes and suffixes. This might help your student remember the names.

Tell your student that the scalene triangle shown on p. 126 is also a right triangle since it has one right angle. (Scalene triangles do not have to be right triangles, as will be seen in the first task.) Ask her if a right triangle could be isosceles (yes) or equilateral (no).

> equi-: equal lateral: sides
> Equilateral triangle: triangle with equal sides
>
> Iso-: equal -skelos: legs
> Isosceles triangle: triangle with "equal legs"
>
> Scalenus: unequal
> Scalene triangle: triangle with unequal sides

Task 1, p. 126

Remind your student that the two triangles with right angles are also called *right triangles*. The one with an obtuse angle is called an *obtuse triangle*. All the other triangles have only acute angles. They are called *acute triangles*.

> 1. **B** and **F** are scalene.
> **A**, **C**, **D**, and **E** are isosceles.
> **C** and **E** are equilateral.
> **B** and **D** have a right angle.
> **F** has an obtuse angle.
> **A** triangle cannot have two obtuse angles.

Practice

Task 2, p. 127

> 2. AC = 5 cm FE = 3 ft
>
> GH = (22 in. – 4 in.) ÷ 2
> = 18 in. ÷ 2
> = **9 in.**
>
> MK= (30 cm ÷ 3)
> = **10 cm**

Workbook

Exercise 10, pp. 142-143 (answers p. 145)

Reinforcement

Extra Practice, Unit 4, Exercise 6, pp. 75-76

Test

Tests, Unit 4, 6A and 6B, pp. 133-138

Chapter 7 – Circles

Objectives

- Identify the center, diameter, and radius of a circle.
- Measure the radius or diameter of a circle.
- Find the radius of a circle given its diameter.
- Find the diameter of a circle given its radius.

Vocabulary

- Circle
- Diameter
- Radius
- Radii

Material

- String, thumbtack, cardboard
- Paper circle
- Ruler
- Compass

Notes

In this chapter, your student will learn to recognize the center, radius, and diameter of a circle.

A **circle** is a set of points which are all the same distance from a given point, the center. The center is usually labeled with the letter "O" in this curriculum.

A **radius** is any line segment from the center of the circle to a point on the circle. A circle has an infinite number of **radii** (plural for radius), all the same length. The term *radius* is also used to mean the *length* of the radius.

A **diameter** is a line segment that has its endpoints on the circle and passes through the center of the circle. The term *diameter* is also used to mean the *length* of the diameter.

The diameter of a circle is twice its radius.

In *Primary Mathematics* 6, students will learn how to find the circumference (perimeter) and area of a circle, using the diameter or the radius.

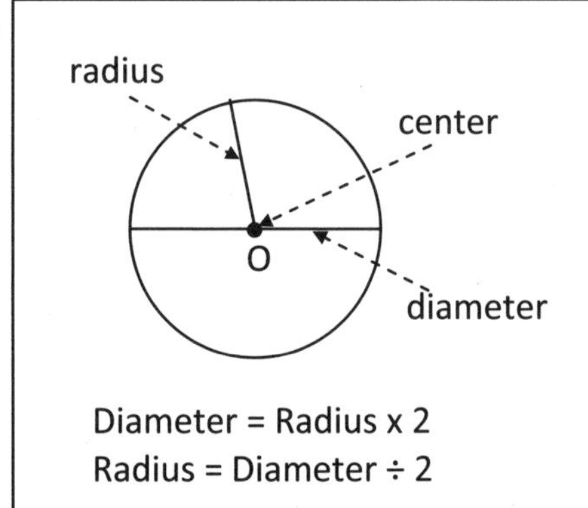

Diameter = Radius x 2
Radius = Diameter ÷ 2

(1) Identify and measure the diameter and radius of a circle

Activity

You may want to do the following activity to give your student an understanding of why all diameters and radii of a circle are equal.

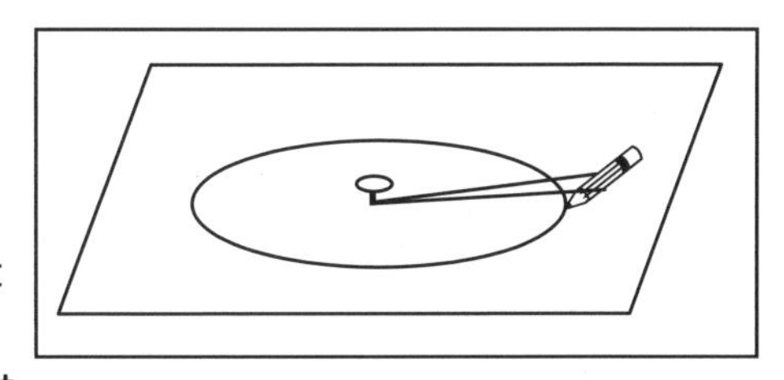

Use a piece of cardboard, a thumbtack, and some string. Place a piece of paper on the cardboard and stick a tack through the paper into the cardboard. Tie the string in a loop so that the doubled length is less than the distance from the tack to the edge of the paper. Have your student loop the string around the tack and the pencil and draw a circle, keeping the string tight and trying to draw a straight line. The string pulls the pencil around in a circle. This can be tricky to do, so you might have to demonstrate for him instead.

Point out that the distance from the tack to the line that is drawn is always the same.

You can also show your student how to use a compass and let her practice drawing circles. The distance from the center of the circle to any point on the circle is always the same.

Discussion

Concept p. 128

In this activity, the diameter and radius are introduced concretely through folding a paper circle. If the circle is folded carefully in half and in half again to make fourths, the intersection of the two creases is the center of the circle. Have your student do this activity using a paper circle.

You can also have your student fold the circle in half and quarters different ways to see that the different diameters and radii are all the same length.

Task 1, p. 129

Be sure that your student realizes that the diameter must go through the center of the circle. So RS cannot be a diameter.

1. (a) **MN** and **PQ** are diameters.
   (b) **RS** is not a diameter because it does not go through the center.
   (c) 12 cm
   (d) 4 in.

Practice

Task 2, p. 129

2. (a) 12 in.
   (b) 5 cm
   (c) $3\frac{1}{2}$ ft

Workbook

Exercise 11, pp. 144-145 (answers p. 145)

Reinforcement

Extra Practice, Unit 4, Exercise 7, pp. 77-78

Test

Tests, Unit 4, 7A and 7B, pp. 139-142

Chapter 8 – Solid Figures

Objectives

- Visualize cubes, prisms, pyramids, and cylinders from two-dimensional drawings.
- Determine the number and shapes of the faces of a solid from a two-dimensional drawing.

Vocabulary

- Prism
- Pyramid
- Cylinder
- Vertex
- Base
- Apex

Material

- Solid models of prisms, pyramids, and cylinders

Notes

In *Primary Mathematics* 2B, students learned to identify some basic solids: cubes, prisms, rectangular prisms, pyramids, cones, and spheres. This chapter is a review of prisms, pyramids, and cylinders, with an emphasis on determining the number of faces and vertices for the prisms and pyramids from a two-dimensional drawing of the solid. In *Primary Mathematics* 2B, students identified faces and edges that were curved or flat. In *Primary Mathematics* 3B, the term face was restricted to a flat surface. However, in *Primary Mathematics* the definition of a face is not restricted to a polygon, as it will be later in a formal study of geometry, and an edge, where two faces meet, does not have to be a straight line. A **vertex** is where three or more edges meet.

A polyhedron is a three-dimensional solid which consists of a collection of polygons, joined at their edges. (Students are not required to learn the term polyhedron here.)

A general **prism** is a polyhedron possessing two congruent polygonal faces, which we will call **bases**, with all remaining faces parallelograms. At this level, students will only deal with right prisms in which the top and bottom polygons lie on top of each other so that the vertical polygons connecting their sides are not only parallelograms, but also rectangles.

A **pyramid** is a polyhedron where one face (the base) is a polygon and all the other faces are triangles meeting at a common vertex, or **apex**. A right pyramid is a pyramid for which the line joining the centroid of the base and the apex is perpendicular to the base. A regular pyramid is a right pyramid whose base is a regular polygon.

Prisms and pyramids can be named by the shape of the bases; for example, a prism with rectangular bases is called a rectangular prism and a pyramid with a triangular base is called a triangular pyramid.

A **cylinder** has two circles as bases.

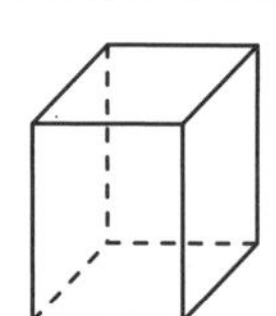
rectangular prism

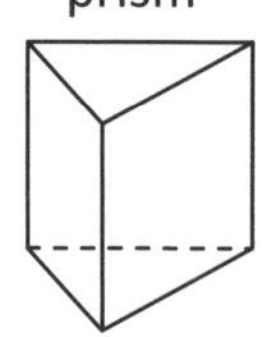
triangular prism

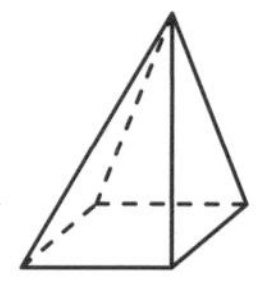
rectangular pyramid

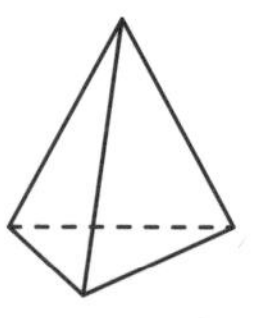
triangular pyramid

cylinder

(1) Identify the faces and vertices of prisms and pyramids

Discussion

Concept p. 130

Relate the drawings on this page to actual physical solids, if necessary. Discuss the difference between a prism and a pyramid. Be sure your student understands that a *prism* has two polygonal faces that are the exact same shape and size, which we can call *bases*. The remaining faces are all parallelograms. If the bases are directly on top of each other, as in the examples in the text, the remaining faces are rectangles. A *pyramid* has a polygon for one base, and all the other faces are triangles that meet at a common *vertex*, also called an *apex*.

Ask your student how the first part of the names (rectangular, triangular, square) were determined. The prisms are named by the shape of the bases. We could have octagonal prisms or pentagonal pyramids. Remind her that a cube is a special type of rectangular prism where all the faces are squares.

	Faces	Edges	Vertices
Rectangular prism	6 (2 rectangles for bases, 4 rectangles for sides)	12	8
Triangular prism	5 (2 triangles for bases, 3 rectangles for sides)	9	6
Square pyramid	5 (1 square for base, 4 triangles for sides)	8	5
Triangular pyramid	4 (1 triangle for base 3 triangles for sides)	6	4

Get your student to identify the types of faces as well as count them, along with counting the vertices and edges.

Discuss ways in which the shape of the base can determine the number of faces, edges, and vertices. In the table at the right, B stands for the number of sides on the base (e.g. 4 for rectangular prism or pyramid, 3 for triangular prism and pyramid).

	Faces	Edges	Vertices
prism	B + 2	B x 3	B x 2
pyramid	B + 1	B x 2	B + 1

Practice

Task 1, p. 131

1. A cylinder has 1 curved side and 2 faces.
2. Triangular prism: 2 triangular faces, 3 rectangular faces.
   Triangular pyramid: 4 triangular faces.
   Rectangular pyramid: 4 triangular faces, 1 rectangular face.
3. D is different since it is a pyramid and the rest are prisms.

Workbook

Exercise 12, pp. 146-147 (answers p. 145)

Reinforcement

Extra Practice, Unit 4, Exercise 8, pp. 79-80

Test

Tests, Unit 4, 8A and 8B, pp. 143-148

Chapter 9 – Nets

Objectives

- ♦ Form solids from nets.
- ♦ Identify nets of cubes.
- ♦ Identify the solid represented by a net.

Vocabulary

- ♦ Net

Material

- ♦ Square graph paper (appendix p. a17)

Notes

In this chapter, your student will learn about nets of sold figures.

A **net** of a solid is a two-dimensional figure which can be folded to form the surface of the solid. The drawing at the right shows the net of a triangular pyramid and a cube. A solid can have more than one net.

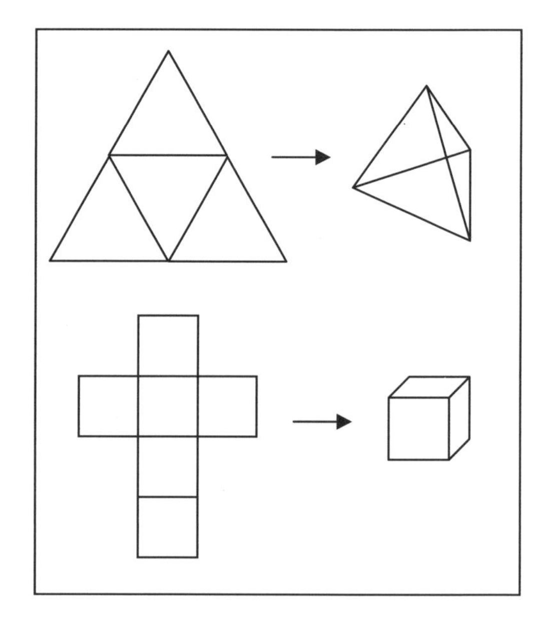

Being able to visualize the net of a solid will help students find the surface area of solids later in *Primary Mathematics* 6B.

Identifying the solid formed by a given net, or a net that could be formed from a given solid, can be difficult for some students. If your student has difficulty, have him trace the nets in the learning tasks and the workbook exercises and try folding each of them into a solid to see concretely why they may or may not form the solid.

Even without tracing and cutting out the shapes, some choices can be easy to rule out. The nets need to have the same number and type of faces as the faces of the solid. From the previous chapter, your student should be able to determine the number and type of faces on any of the solids she will encounter in this chapter. Any point on the net where three or four lines come together is going to be the vertex of the solid. Since adjacent edges of the nets will go together to form an edge, they must be the same length as the edges on the solid.

(1) Form solids from nets

Activity

If you have any boxes available, you can have your student examine them and perhaps pull apart the glued parts to see how the boxes are all made from flat sheets of cardboard. (These flat sheets are not really nets, since there is overlap where the sides are glued together and any other possible overlapping flaps for closing the top and bottom.)

Discussion

Concept p. 132

Page 132 introduces nets. Have your student do the activity on this page. Tell him that the net shows all the faces of the solid laid out flat.

After your student has formed the rectangular prism, help her to tape the edges together, and then cut apart several other edges to form a different net. Have her compare the new net to the one on p. 132. A solid can have more than one net.

Point out that a net contains the whole faces of the solid, as if the solid is cut only along an edge. If the paper solid were cut across a face to flatten it out, the resulting figure would not be a net.

Tasks 1-2, pp. 133-134

1: Have your student name the shape formed from the net.

2: You can have your student copy the nets onto square graph paper, cut them out, and see if they fold into a cube. All of them can form cubes.

1. (a) triangular prism

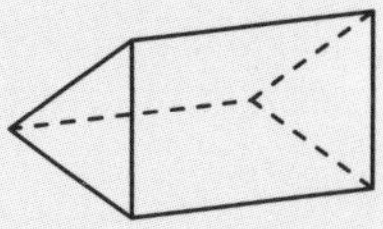

(b) rectangular or square pyramid

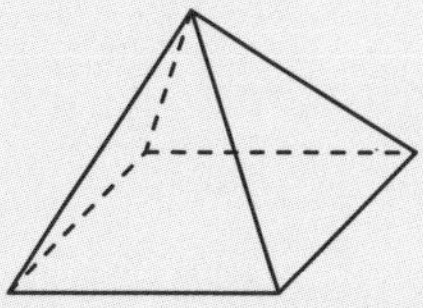

2. All of them can form a cube.

Workbook

Exercise 13, pp. 148-149 (answers p. 145)

Enrichment

See if your student can come up with any other nets that will form a cube. There are 11 possible nets, and they are all represented on p. 134 of the textbook. There are some general rules that can be used to determine whether a joined set of 6 squares can form a cube. These rules could also be used to determine if a net can form a rectangular prism:

⇒ The longest panel cannot have more than 4 squares.

⇒ If the longest panel has 4 squares, the remaining 2 have to be placed on opposite sides.

⇒ If the longest panel has 3 squares, the remaining 3 can be placed on either side but there must always be at least 2 free edges on either side of the longest panel.

⇒ If the longest panel has 2 squares, the remaining 4 must be evenly distributed on either side such that there will be one free edge on either side of the panel.

(2) Identify nets of a solid and solids of a net

Discussion

Task 3-4, pp. 135-136

> 3. A and D
>
> 4. B

3: In this task, your student must decide which of the four nets are nets of the solid. See if he can find the answer without tracing the nets.

The nets need to have the same number and type of shapes as the faces of the solid. So B is obviously not a net of the solid because it would have only 3 faces. Any point on the net where three or four lines come together is going to be a vertex of the solid.

Adjacent edges of the nets must be the same length, as they will go together to form an edge. So C could not be a net because one of the triangles that would fold up from the center triangle has a longer edge than the one that would be right next to it.

(In the 2008 printing, none of the nets would form the solid. A and D have been distorted.)

4: In this task your student will see a net and several 2-dimensional drawings of solids, and will have to choose which of the solids could be formed from the net. Again, see if she can find the answer without tracing the net.

The net has 5 faces, so neither A nor C could be the solid, since A has 6 faces and C has 4. The net has two triangles, so D cannot be the solid, since it has 4 triangular faces. B has the correct number of rectangular and triangular faces.

Workbook

Exercises 14-15, pp. 150-155 (answers p. 145)

Reinforcement

Extra Practice, Unit 4, Exercise 9, pp. 81-82

Test

Tests, Unit 4, 9A and 9B, pp. 149-156

Workbook

Exercise 9, pp. 140-141

1. (a) QR = 5 cm — QR // **PS**
   RS = 3 cm — PS // **QR**
   PS = **5 cm** — PQ ⊥ **PS**
   PQ = **3 cm** — and **QR**
   (b) XY = 10 m
   YZ = 4 m — WZ // **XY**
   WZ = **10 m** — WX // **ZY**
   WX = **4 m**
   (c) BC = 8 cm — AB // **DC**
   AB = **8 cm** — BC // **AD**
   CD = **8 cm** — AB ⊥ **AD**
   AD = **8 cm** — and **BC**
   (d) MN = 3 m
   LM = **3 m** — ML // **NK**
   LK = **3 m** — LK // **MN**
   NK = **3 m**

2. (a) EB = **3 cm**
   FE = **5 cm** (8 cm – 3 cm)
   (b) JK = **5 cm** (11 cm – 6 cm)
   MH = **9 cm** (6 cm + 3 cm)
   (c) TS = **4 m** (11 m – 7 m)
   XS = **17 m** (6 m + 11 m)

Exercise 10, pp. 142-143

1. (a) true
   (b) false
   (c) false
   (d) false
   (e) true
   (f) true

2. (a) QR = **8 in.**
   PQ = **8 in.**
   (b) MO = **8 yd**

3. (a) scalene (b) isosceles (c) equilateral
   (d) scalene (e) equilateral

4. isosceles

Exercise 11, pp. 144-145

1. (a) 40 in.
   (b) 34 yd

2. (a) 32 ft
   (b) 64 in.

3. AD

4. (a) 10 cm
   (b) 5 cm
   (c) 5 cm

Exercise 12, pp. 146-147

1. Rectangular prism ⟶ apple juice carton
   Triangular pyramid ⟶ tent
   Triangular prism ⟶ cake slice
   Cylinder ⟶ can

2.

	Faces	Edges	Vertices
Rectangular prism	6	12	8
Triangular prism	5	9	6
Square pyramid	5	8	5
Triangular pyramid	4	6	4

Exercise 13, pp. 148-149

1. A, C, F, G, H
2. B, C, E, G, H

Exercise 14, pp. 150-152

1. A, C
2. A, D
3. A, D

Exercise 15, pp. 153-155

1. C
2. C
3. C

Review 4

Review

Review 4, pp. 137-140

(Note: Problems 10(a) and 10(b) in the 2008 printing are inappropriate. Add a 0 to 100 to make 1000 for both.)

Workbook

Review 4 pp. 156-161 (answers p. 147)

Test

Tests, Units 1-4, Cumulative A and B, pp. 157-167

1. (a) > (b) <
   (c) > (d) >
   (e) = (f) >
   (g) > (h) >

2. e

3. (a) the center of the circle
   (b) radius
   (c) diameter
   (d) 8 in.

4. 10 cm and 8 cm

5. square

6. 5 faces; triangular prism

7. (a) A: 300,000 B: 10,000 C: 150,000 D: 230,000
   (b) A: 0 B: –20 C: –8 D: 16

8. A: $\frac{1}{2}$ B: $1\frac{2}{5}$ C: $2\frac{7}{10}$ D: $4\frac{4}{5}$

9. 0, –3, –5, 4

10. (a) 250 (b) 350
    (c) 5450 (d) 7200
    (e) 528 (f) 184
    (g) 21 (h) 2136

11. (a) $n = \frac{3}{7}$ (b) $n = \frac{2}{9}$
    (c) $n = \frac{7}{10}$ (d) $n = \frac{11}{12}$

12. (a) (24 x 18) – 100 = 332 or 24 x 18 – 100 = 332
    (b) (1468 + 602) ÷ 6 = 345

13. $\frac{9}{6} = \frac{3}{2} = \mathbf{1\frac{1}{2}}$

14. 48

15. 6

16. (a) $\frac{5}{12}, \frac{1}{2}, \frac{3}{4}, 1\frac{1}{2}$
    (b) $\frac{2}{5}, \frac{6}{15}, \frac{7}{10}, 1$ or $\frac{6}{15}, \frac{2}{5}, \frac{7}{10}, 1$

17. $\frac{40 \text{ in.}}{6} = \mathbf{6\frac{2}{3} \text{ in.}}$ Each piece is $6\frac{2}{3}$ in. long.

18. $\frac{3}{10}$ of \$40 = 3 x \$$\frac{40}{10}$ = 3 x \$4 = **\$12**
    The book cost \$12.

19. $1 - \frac{1}{3} - \frac{4}{9} = 1 - \frac{3}{9} - \frac{4}{9} = \mathbf{\frac{2}{9}}$
    He traveled $\frac{2}{9}$ of the trip the third day.

20. ?
    8
    1 unit = 8
    3 units = 8 x 3 = **24**
    He bought 24 toy cars.

21. $1 - \frac{2}{5} = \frac{3}{5}$
    $\frac{3}{5}$ of 100 = 3 x $\frac{100}{5}$ = 3 x 20 = **60**
    There are 60 orange trees.

22. 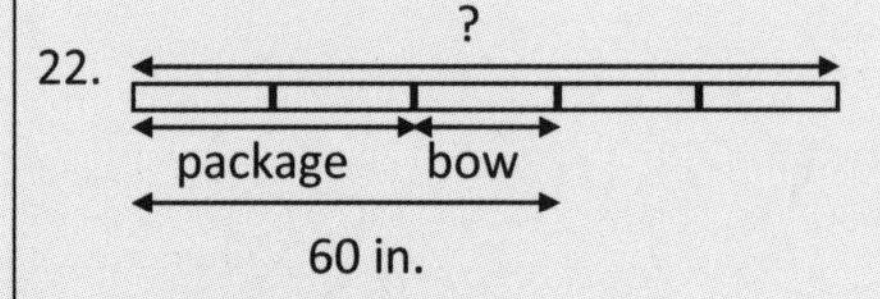

    3 units = 60 in.
    1 unit = 60 in. ÷ 3 = 20 in.
    5 units = 20 in. x 5 = **100 in.**
    She bought 100 in. of ribbon.

Workbook

Review 4, pp. 156-161

1. 56,952
2. 85,320
3. 76,410
4. 6
5. 8, 16, 24, 32, 40
6. 15
7. 130 cm
8. 459 x 24 = 11,016 → **11,000**
9. (a) <
   (b) >
   (c) <
10. $\frac{3}{8}$
11. A: $3\frac{1}{4}$ B: $3\frac{7}{8}$
12. \$1857 ÷ 3 = **\$619**
    The printer costs \$619.
13. \$1460 x 6 = **\$8760**
    He earns \$8760 in 6 months.
14. 64 cm ÷ 4 = **16 cm**
    Each side is 16 cm long.
15. $\frac{2}{5}$ (This is the only one that is less than a half.)
16. $\frac{2}{9}, \frac{2}{7}, \frac{2}{3}, \frac{9}{7}$
17. $1\frac{5}{8}; 1\frac{3}{4}; 2\frac{1}{8}$
18. 6 out of 10 = $\mathbf{\frac{3}{5}}$
19. $\frac{13}{5}$
20. $4\frac{3}{4}$
21. $\frac{3}{5} \times 10 = \frac{3}{5}$ of $10 = 3 \times \frac{10}{5} = 3 \times 2 = \mathbf{6}$
22. $\frac{3}{5}$ yd + $\frac{3}{5}$ yd + $\frac{3}{5}$ yd + $\frac{3}{5}$ yd + $\frac{3}{5}$ yd + $\frac{3}{5}$ yd
    = $\frac{18}{5}$ yd = $\mathbf{3\frac{3}{5}}$ **yd** She needs $3\frac{3}{5}$ yards of string.
23. $\frac{3}{5}$ of $50 = 3 \times \frac{50}{5} = 3 \times 10 = \mathbf{30}$
    She made 30 vegetarian pizzas.
24. $\frac{2}{8} = \mathbf{\frac{1}{4}}$
    Each girl gets $\frac{1}{4}$ of a cake.
25. $\frac{3}{4}$ kg + $\frac{1}{2}$ kg + 1 kg
    = $\frac{3}{4}$ kg + $\frac{2}{4}$ kg + 1 kg
    = $\mathbf{2\frac{1}{4}}$ **kg**
    She bought $2\frac{1}{4}$ kg of vegetables.
26. 9:45 p.m.
27.

✓	✓	×	✓
✓	×	×	×
×	×	✓	×
✓	×	×	✓
×	×	✓	×
✓	✓	×	×
×	×	×	×

28. (a) 70 cm (b) 48 m
29. (a) 10 cm (b) 5 cm
30. (a) 30 x (40 + 50) — The price of one...
    (b) 30 + 40 x 50 — The price of forty ...
    (c) 30 x 40 + 50 — The price of forty...
31. 1185 ÷ 3 = 395
    He made 395 sticks of beef kebab.
32. 35 x 150 = **5250**
    He sold 5250 biscuits.
33. 6 yd
    (a) 3 units = 6 yd
    1 unit = 6 yd ÷ 3 = 2 yd
    5 units = 2 yd x 5 = **10 yd**
    She used 10 yd for the bows.
    (b) $\frac{10 \text{ yd}}{6} = 1\frac{4}{6}$ yd = $\mathbf{1\frac{2}{3}}$ **yd**
    Each bow was $1\frac{2}{3}$ yd long.

Unit 5 – Area and Perimeter

Chapter 1 – Area of Rectangles

Objectives

- Review area of polygons.
- Understand and use the formula for area of a rectangle.

Vocabulary

- Area
- Formula

Material

- Centimeter graph paper (appendix p. a17)
- Ruler, yard stick, meter stick

Notes

In *Primary Mathematics* 2B, students were introduced to the concept of area and found the area of polygons by counting square units. In *Primary Mathematics* 3B, they found the area using standard units, such as square centimeter or square inch.

The **area** of a surface is the amount of material needed to "cover" it completely. At this level, your student will only deal with flat surfaces. Area is measured in square units. A square unit is any square whose sides are 1 unit long. If the area of a figure is 4 square centimeters, then it covers the same amount of flat space as would four squares with 1-centimeter sides.The units can be any of the standard units. So area can be measured in square centimeters, square meters, square feet, square yards, square miles, and so on.

In *Primary Mathematics* 3B, students used the terms square centimeter or square inch for the unit. In this chapter, your student will learn to represent the unit with the superscript 2, e.g. cm^2 or $in.^2$

In this chapter, your student will find the area of rectangles and derive the formula for the area of a rectangle: Area = length x width. He will also learn to use mathematical formulas, where a letter is used to stand for specific values. The letters are then replaced with the information from the problem.

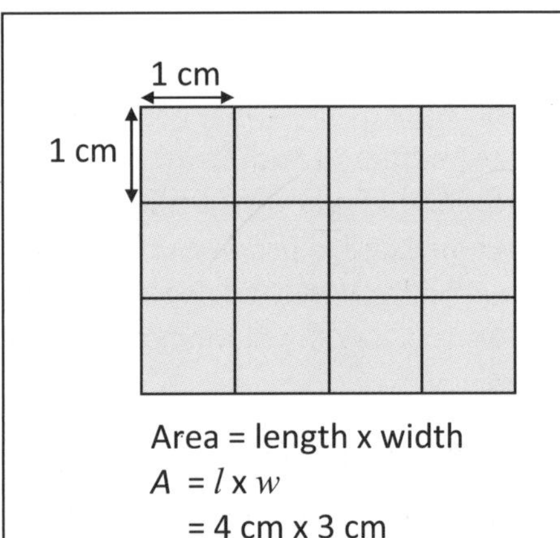

Your student should not simply memorize mathematical formulas. She needs to understand how they were derived so that she can then derive them again herself, and can better understand how they can be manipulated. In the next chapter, she will be finding the length or width given the area and the length of a side.

(1) Find the area of a rectangle

Activity

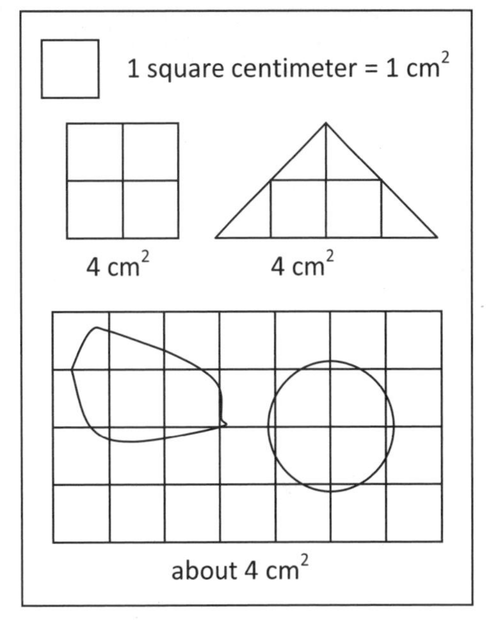

Use centimeter graph paper. Remind your student that the *area* of a figure is the amount of flat space it covers. Area is measured in square units. Outline a square on the centimeter graph paper and ask him for the area. Tell him that we can write square centimeter as cm^2. The little 2 tells us that the measurement unit has two directions at right angles to each other.

Tell your student that if the area of the figure is 4 cm^2, then it covers the same amount of space as four squares, each of whose sides measures 1 cm. Draw a square showing 4 square units and another figure, such as a triangle, which also has an area of 4 square units. Tell her that both of these have the same area, even though they have different shapes. The shape does not have to be squares or half-squares; an irregular shape or circle could have the same area.

Remind your student that we can use other units of area, such as square inch, square foot, or square meter. You can use a ruler, yard stick, and meter stick to give him an idea of the size of a square inch, square foot, square yard, or square meter. You may want to discuss some other units of area. In the metric system, a square kilometer is a square with 1000 meters on each side. An *are* is a square with 10 meters on each side. A *hectare* is a square with 100 meters on each side. In U.S. customary measurement, an *acre* can be covered up with 4840 square yards, and a square mile can be covered up with 640 acres.

Discussion

Concept pp. 141-142

A: 8 cm^2
B: 15 cm^2
C: 21 cm^2

In this activity, your student will look at rectangles drawn on square graph paper in order to derive the formula for the area of the rectangle. Students have used arrays of squares in earlier levels to understand the commutative property of multiplication, and so should not have much difficulty with this concept. Be sure your student understands that we are multiplying the number of rows of squares by the number of columns of squares.

Tasks 1-2, p. 142

1. 20 cm^2

2. 1 $in.^2$

$A = l \times w$
$= 5 \text{ cm} \times 4 \text{ cm}$
$= 20 \text{ cm}^2$

1: Write the formula at the top of the page and show how we can substitute the information given in the problem for the letters in the formula. Remind your student that it does not matter what order we put the factors in, so it is just as valid to write that A = w x l = 4 cm x 5 cm, but usually the formula is written with length first.

2: For a square, length = width.

Task 3, p. 143

3. (a) square feet
   (b) square miles
   (c) 3 feet
   (d) No. 1 $yd^2 = 9\ ft^2$
   (e) 1000 m
   (f) 10,000,000,000 cm^2

Tasks 3(a)-3(b), and the information in the green talk-bubble at the bottom of p. 142, are meant to remind your student that we can use other units of area than square centimeters.

Tasks 3(c)-3(f) are meant to remind your student that when we are given the length of the sides in one measurement unit and asked to find the area in a smaller unit, we need to convert the sides to the unit for the area before we multiply to find the area; if the area is to be in square feet, then the sides must be in feet. A square 1 yard on the side is a square yard, but 1 square yard is not the same as 3 square feet, even though a linear yard equals 3 linear feet. If necessary, you can draw a diagram to illustrate this concept.

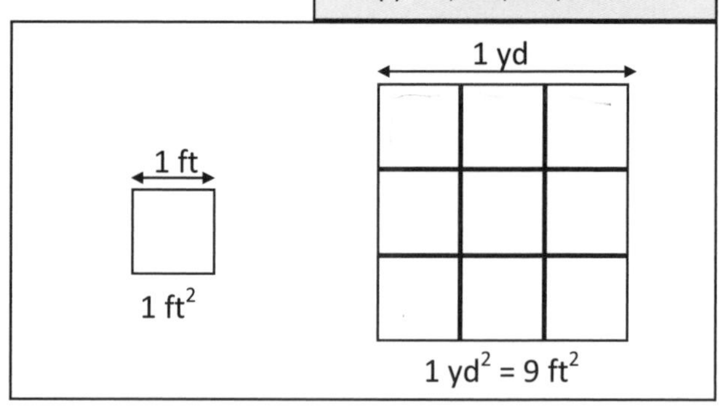

To find the answer to 3(f), your student could first find the number of square meters in a square kilometer, then the number of square centimeters in a square meter, and multiply that by the number of square meters in a kilometer. Or she can first find the number of centimeters in a kilometer. The answer is ten thousand million (or 10 billion; students will learn about billions in *Primary Mathematics* 5A).

1 km = 1000 m
1 km^2 = 1000 m x 1000 m = 1,000,000 m^2
1 m = 100 cm
1 m^2 = 100 cm x 100 cm = 10,000 cm^2
1 km^2 = 10,000 x 1,000,000 = 10,000,000,000 cm^2
Or
1 km = 1000 m
1 m = 100 cm
1 km = 100 x 1000 m = 100,000 cm
1 km^2 = 100,000 x 100,000 = 10,000,000,000 cm^2

Practice

Task 4, p. 143

4. (a) 12 cm^2
   (b) 18 cm^2

Practice A, p. 144

(Note: In the 2008 printing, problems 1 and 3 of Practice A involve perimeter, but your student should be able to answer it from concepts learned in earlier levels. If necessary, tell your student that perimeter is the distance around and object. You can also save this practice to do along with Practice B.)

1. A: 25 cm^2 B: 374 $in.^2$
   C: 432 m^2
   D: 64 ft^2 E: 300 mi^2
2. 150 cm^2
3. 190 m

Workbook

Exercise 1, pp. 162-164 (answers p. 156)

Reinforcement and Test

The 2008 printings of *Extra Practice* and *Tests* have some inappropriate problems for this chapter. Save them to use in the next chapter.

Chapter 2 – Perimeter of Rectangles

Objectives

- Review perimeter of polygons.
- Understand and use formulas for perimeter of a rectangle.
- Solve problems involving finding an unknown side of a rectangle, given either the area or the perimeter and the length of one side.

Vocabulary

- Perimeter

Material

- Square grid paper (appendix p. a17)

Notes

In *Primary Mathematics* 3B, students were formally introduced to the concept of perimeter. They had already been finding the perimeter informally, as in finding the distance around a triangle or a track in the units on length.

In this chapter, your student will review the concept of perimeter, and derive some formulas for the perimeter of a rectangle.

The **perimeter** of a figure is the distance around the outside of the figure.

To find the perimeter of a rectilinear figure drawn with squares, we can add all the lengths along the outside. The figure at the right has a perimeter of 14 cm.

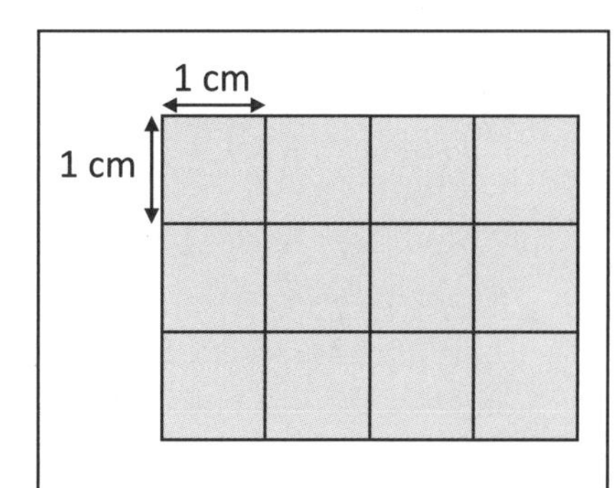

Perimeter = length + width + length + width

$P = l + w + l + w$

$= 2 \times l + 2 \times w$

Or

$P = 2 \times (l + w)$

$A = l \times w$

$l = A \div w$

$w = A \div l$

$P = 2 \times (l + w)$

$l = (P \div 2) - w$

$w = (P \div 2) - l$

Since pairs of sides of a rectangle are equal in length, we can find the perimeter by doubling two of the adjacent sides, either before or after adding them together. (The two formulas give students an intuitive understanding of the distributive property of multiplication, which they will learn formally in *Primary Mathematics* 5A.)

Your student will also find the length of a side given the perimeter using the formulas for perimeter and knowledge of the relationship between multiplication and division.

A square is a special rectangle where the length and the width are the same. Your student will be asked to find a side of a square when given its area. He should be able to recognize square numbers through 10 x 10 so that he can find the side of a square given its area (side × side). In *Primary Mathematics* 5A, students will learn how to find the prime factorization of a number, which can be used to find the square root of a number.

The term "square root" and the symbol for square root ($\sqrt{\ }$) are not used at this level.

(1) Find the perimeter of a rectangle

Discussion

Concept p. 145

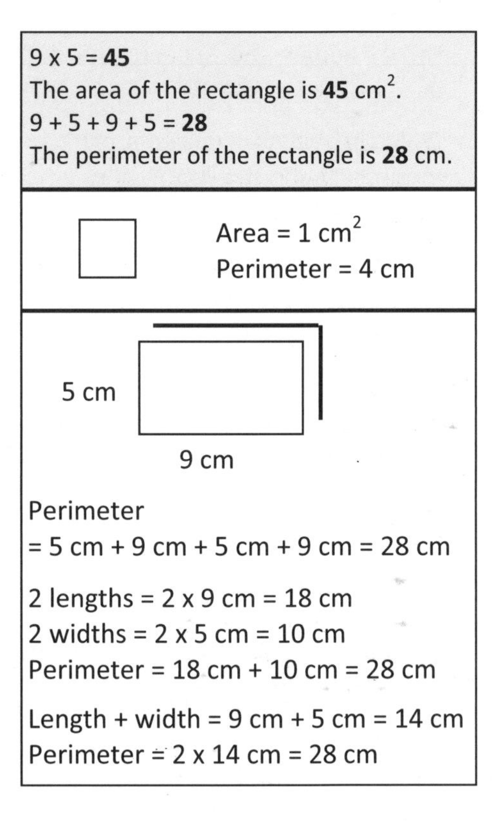

After your student has found the area, remind her that the distance around a figure is its *perimeter*. Draw a single square centimeter or mark one on centimeter graph paper and ask her to find the perimeter. Each side has to be counted, so the perimeter is 4 cm. Then have her find the perimeter of the rectangle on p. 145. We can count the centimeters all the way around the sides until we get back to the starting point.

If, however, the squares are not marked, then we will be told the length and width of the sides. Draw a rectangle and label the length and width **9 cm** and **5 cm** respectively. We can find the perimeter by simply adding the lengths of each side all the way around.

Ask your student for other ways to calculate the perimeter. Since both lengths and widths are the same, we can find the perimeter by doubling the length and width, and then adding them together. Or, since there are two lengths and widths, we can add one length and one width, and then double the sum.

Have your student look at the formulas at the bottom of page 145. Ask him which one he prefers. The last one involves the fewest steps, but the second one can be used if it makes mental calculations easier, such as in this example where doubling the width gives 10 cm.

Practice

Tasks 1-2, p. 146

1. (a) Area = 96 cm^2
   Perimeter = 40 cm
   (b) Area = 160 m^2
   Perimeter = 56 m
2. Area = 25 m^2
   Perimeter = 20 m

Workbook

Exercise 2, pp. 165-168 (answers p. 156)

(2) Find the length of a side of a rectangle

Activity

Area = 100 cm^2
Side = 10 cm

Draw a square and tell your student that the area is 100 cm^2. Ask her to find the length of the side.

Since the area of a square = side x side, we need to find a number which, when multiplied by itself, gives 100. That number is 10, so the side of the square is 10 cm.

Tell your student that 100 is a square number. A square number is the product you get when you multiply one number by itself. Any area of a square is a square number.

Get your student to list all the square numbers through the product of 12 x 12. You may want to tell him that a dozen dozen, or 144, is sometimes called a "gross" (a word that now means something different to most students this age!).

Provide your student with some practice finding the square root of square numbers. You could make a set of cards with the square numbers on them, mix them up, and have her go through them periodically, telling you the number that, when multiplied by itself, gives the number on the card.

$1 \times 1 = 1$
$2 \times 2 = 4$
$3 \times 3 = 9$
$4 \times 4 = 16$
$5 \times 5 = 25$
$6 \times 6 = 36$
$7 \times 7 = 49$
$8 \times 8 = 64$
$9 \times 9 = 81$
$10 \times 10 = 100$
$11 \times 11 = 121$
$12 \times 12 = 144$

Draw a rectangle and tell your student that the area is 24 cm^2. Ask him if we can find the sides. Since they are not equal, we do not have enough information to find either the length or the width. However, if we are given one side, we can find the other. Tell him that the width is 4 cm and ask him to find the length. This is the same as problems he has had earlier with a missing factor. To find the missing factor, we divide by the given factor. So to find the length, we divide the area by the width.

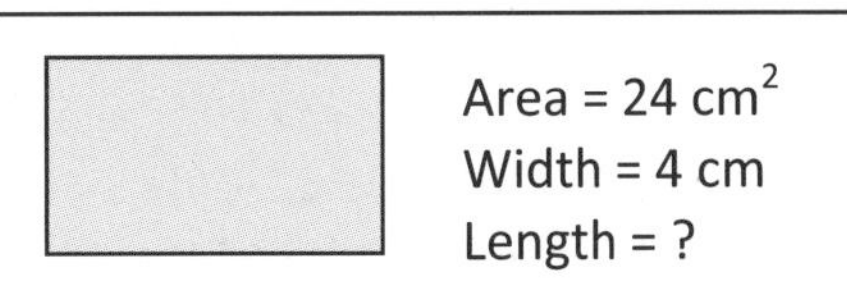

Area = 24 cm^2
Width = 4 cm
Length = ?

$A = l \times w$
$24\ cm^2 = l \times 4\ cm$
$l = 24\ cm^2 \div 4\ cm = 6\ cm$

Discussion

Tasks 3-4, pp. 146-147

3. 3; 3 cm

4. 4 m

3: Since the sides of a square are all equal, all we need to do is divide the perimeter by 4.

4: The four sides of the rectangle are not equal, but each length and width are equal to the other. So we can find the (length + width) by dividing the perimeter by 2. If we are then given either the length or the width, we can find the other side by subtracting the known side from half the perimeter. Ask your student if she can come up with a different method. In this task, she could subtract 8 twice (16) from the perimeter. Then divide the answer by 2 to get the width.

Practice

Tasks 5-8, pp. 147-148

These tasks involve problems that combine area and perimeter. See if your student can solve them independently first.

5. (a) Perimeter of square = 10 cm x 4 = 40 cm
   Perimeter of rectangle = 40 cm
   $l + w$ = 40 cm ÷ 2 = 20 cm
   l = 20 cm – 5 cm = **15 cm**
   The rectangle is 15 cm long.
   (b) Area of rectangle = 15 cm x 5 cm = 75 cm^2
   Area of square = 10 cm x 10 cm = 100 cm^2
   The **square** has a bigger area.

6. Length of one side = **5** m
   Area of square = 5 m x 5 m = **25** m^2

7. Perimeter of square = 6 cm x 4 = **24** cm

8. w = **5**
   Width = **5** m
   Perimeter = **26** m

Workbook

Exercise 3, pp. 169-171 (answers p. 156)

Reinforcement

Extra Practice, Unit 5, Exercise 1, pp. 87-88

Extra Practice, Unit 5, Exercise 2, pp. 89-90

(3) Practice

Practice

Practice B, pp. 149-150

Problem 1: You can point out that a rectangle with a bigger area will not necessarily have a longer perimeter than one with a smaller area.

Problem 9: Students have not yet learned to divide by a 2-digit number, but they have learned the relationship between division and fractions. Your student could write this as a fraction problem that needs to be simplified. Or, simply use mental math, counting up by 15; 15 x 2 = 30, so 15 x 4 = 60, and 15 x 8 = 120.

Test

Tests, Unit 5, 1A and 1B, pp. 169-174

Tests, Unit 5, 2A and 2B, pp. 175-178

1. (a) Width = (40 cm ÷ 2) – 12 cm = **8 cm**
   Area = 12 cm x 8 cm = **96 cm^2**
   (b) Length = (46 cm ÷ 2) – 8 cm = **15 cm**
   Area = 15 cm x 8 cm = **120 cm^2**
   (c) Length = (40 m ÷ 2) – 6 m = **14 m**
   Area = 14 m x 6 m = **84 m^2**
   (d) Width = (52 m ÷ 2) – 15 m = **11 m**
   Area = 15 m x 11 m = **165 m^2**
2. (a) Length = 18 m^2 ÷ 3 m = **6 m**
   (b) Width = 32 m^2 ÷ 8 m = **4 m**
   (c) Length = 96 cm^2 ÷ 6 cm = **16 cm**
   (d) Width = 108 cm^2 ÷ 9 cm = **12 cm**
3. (a) Area of X: 6 in. x 3 in. = 18 in.^{2}
   Area of Y: 8 in. x 2 in. = 16 in.^{2}
   Rectangle **X** has the bigger area.
   (b) Perimeter of X: 18 in.
   Perimeter of Y: 20 in.
   Rectangle **X** has the shorter perimeter.
4. 250 m
5. Area of 1 square = 4 cm^2
   Area of 8 squares = 4 cm^2 x 8 = 32 cm^2
   Or:
   Length = 8 cm
   Width = 4 cm
   Area = 8 cm x 4 cm = **32 cm^2**
   Perimeter = **24 cm**
   The area is 32 cm^2 and the perimeter is 24 cm.
6. Height = 36 m^2 ÷ 9 m = **4 m**
   The wall is 4 m high.
7. Perimeter = 300 m
   1 m: $9
   300 m: $9 x 300 = **$2700**
   It costs $2700 to fence the field.
8. Area = 10 yd^2
   1 yd^2: $6
   10 yd^2: $6 x 10 = **$60**
   It costs $60 to tile the floor.
9. Width = $\frac{120 \text{ mi}^2}{15 \text{ mi}} \overset{\div 5}{=} \frac{24 \text{ mi}^2}{3 \text{ mi}} \overset{\div 3}{=} \frac{8 \text{ mi}^2}{1 \text{ mi}} = 8 \text{ mi}$
   Perimeter = (15 mi x 2) + (8 mi x 2) = 46 mi

Workbook

Exercise 1, pp. 162-164

1.

Rectangle	Length	Width	Area
A	4 cm	2 cm	8 cm^2
B	6 cm	2 cm	12 cm^2
C	7 cm	3 cm	21 cm^2
D	5 cm	3 cm	15 cm^2
E	4 cm	3 cm	12 cm^2

2.

Rectangle	Length	Width	Area
A	5 cm	2 cm	10 cm^2
B	4 cm	3 cm	12 cm^2
C	6 cm	4 cm	24 cm^2
D	7 cm	3 cm	21 cm^2
E	8 cm	1 cm	8 cm^2

3.

Rectangle	A	B	C	D	E	F
Area	12 cm^2	24 $in.^2$	35 m^2	54 yd^2	40 ft^2	120 cm^2

Exercise 2, pp. 165-168

1.

5 cm^2 12 cm	6 cm^2 12 cm
7 cm^2 12 cm	8 cm^2 14 cm
10 cm^2 14 cm	7 cm^2 16 cm

2. Check drawings.

3. (a) The area is **40** cm^2.
   The perimeter is **28** cm.
   (b) The area is **72** mi^2.
   The perimeter is **36** mi.
   (c) The area is **81** km^2.
   The perimeter is **36** km.

4. (Note: the figures are not to scale with each other.)

Figure	Area	Perimeter
A	12 cm^2	16 cm
B	25 cm^2	20 cm
C	24 cm^2	22 cm
D	36 cm^2	24 cm
E	24 cm^2	20 cm

(a) C and E
(b) B and E
(c) 11 cm^2
(d) 2 cm

Exercise 3, pp. 169-171

1. (a) CD = **5 cm**
   Area = 9 cm x 5 cm = **45 cm^2**
   (b) $l + w$ = 42 in. ÷ 2 = 21 in.
   EF = 21 in. – 6 in. = **15 in.**
   Area = 15 in. x 6 in. = **90 $in.^2$**
   (c) $l + w$ = 26 m ÷ 2 = 13 m
   SR = 13 m – 7 m = **6 m**
   Area = 6 m x 7 m = **42 m^2**

2.

Figure	Area	Length	Width	Perimeter
A	48 cm^2	8 cm	**6 cm**	**28 cm**
B	160 ft^2	**16 ft**	10 ft	**52 ft**
C	100 $in.^2$	10 in.	**10 in.**	**40 in.**
D	135 m^2	**15 m**	9 m	**48 m**
E	112 yd^2	**14 yd**	8 yd	**44 yd**

3. Area of 1 stamp: 4 cm x 3 cm = 12 cm^2
   Area of 30 stamps: 12 cm^2 x 30 = **360 cm^2**
   The area of the picture is 360 cm^2.

4. Area of floor: 5 yd x 4 yd = 20 yd^2
   1 yd^2 = \$12
   20 yd^2: \$12 x 20 = **\$240**
   It costs \$240 to carpet the room.

5. Perimeter of field: 140 m
   1 m: \$7
   140 m: \$7 x 140 = **\$980**
   It costs \$980 to put up the fence.

Chapter 3 – Composite Figures

Objectives

- Find the area and perimeter of a figure made up of rectangles and squares.
- Find the area of a path around a rectangle.

Material

- Square grid paper (appendix p. a17)
- Index cards

Notes

In this chapter, your student will use his knowledge of area and perimeter to find the area and perimeter of figures that can be divided into rectangles; that is, one that is *composed* of rectangles.

In *Primary Mathematics* 5, students will find area and perimeter of composite figures that include triangles and parallelograms, and in *Primary Mathematics* 6 they will find the area and perimeter of composite figures that also include circles, semicircles, and quarter circles.

Generally, we find the perimeter by adding the length of all the sides of the figure. We can find the area of the figure itself by dividing it up into rectangles. There are often various ways to divide the figure up into component rectangles. Or, we can find the area of a larger figure and find the area of the figure by subtracting the resulting rectangles.

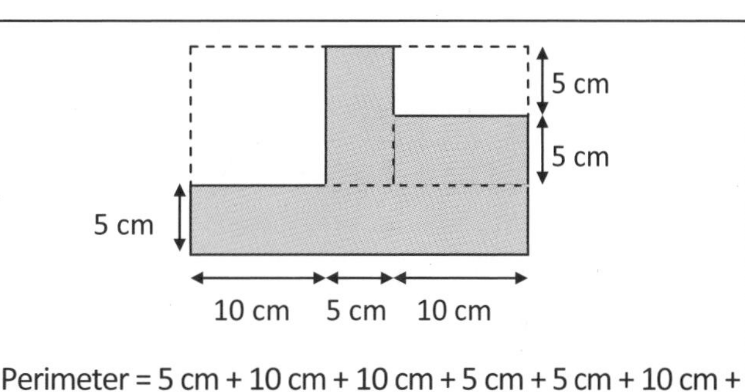

Perimeter = 5 cm + 10 cm + 10 cm + 5 cm + 5 cm + 10 cm + 10 cm + 25 cm = 80 cm
Area = (10 cm x 5 cm) + (10 cm x 5 cm) + (25 cm x 5 cm) = 50 cm^2 + 50 cm^2 + 125 cm^2 = 225 cm^2
Area = (25 cm x 15 cm) – (10 cm x 10 cm) – (10 cm x 5 cm) = 375 cm^2 – 100 cm^2 – 50 cm^2 = 225 cm^2

The idea of subtracting the area of a smaller rectangle(s) from that of a bigger one can be use even if the smaller rectangle is entirely within the larger rectangle. The difference is the area around the smaller rectangle. This is an easier way to find the area of a path around a rectangular figure than dividing the path up into four separate rectangles.

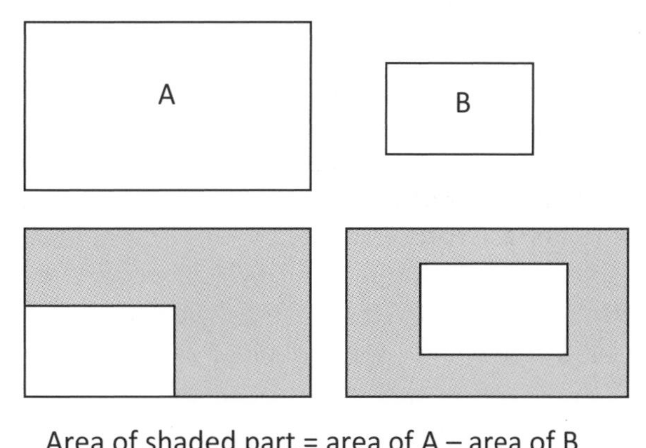

Area of shaded part = area of A – area of B

(1) Find the perimeter of a composite figure

Activity

Cut some index cards into 3 or 4 rectangles with different areas of a whole number of square centimeters, such as 8 cm by 5 cm, 3 cm by 6 cm, and 4 cm by 4 cm. Arrange them on centimeter graph paper such that adjacent rectangles touch at least 1 cm along the sides of both and that they line up with the centimeter squares. You can trace the outline.

Ask your student to find the perimeter and area of the resulting figure. To find the perimeter, she will have to add up the length of all the sides.

Have your student move the rectangles to new positions and find the new area and perimeter. He should realize that the area will stay the same (if the rectangles do not overlap).

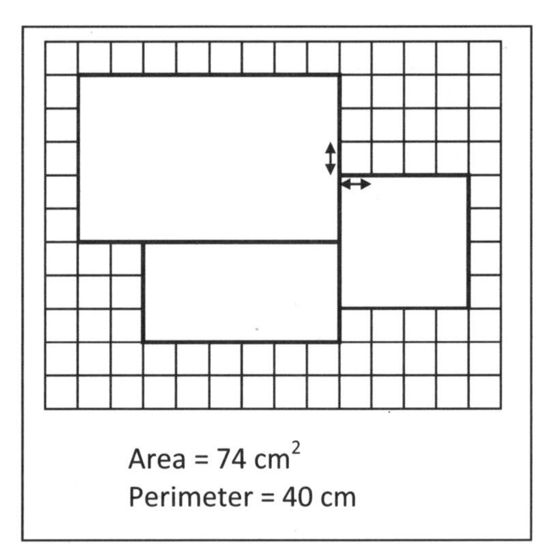

Area = 74 cm^2
Perimeter = 40 cm

Discussion

Concept p. 151

Get your student to find the area by dividing the figures up into two rectangles and using the formula for the area of a rectangle. Point out that each figure can be divided into two rectangles in two different ways. For example, Figure A can be divided into the two rectangles shown by the slightly darker vertical line, which are 5 cm by 2 cm and 3 cm by 4 cm, or by a horizontal line into one rectangle that is 3 cm by 2 cm and another 8 cm by 2 cm.

Tasks 1-2, p. 152

A: Area = 22 cm^2 Perimeter = 24 cm
B: Area = 22 cm^2 Perimeter = 20 cm
C: Area = 22 cm^2 Perimeter = 22 cm
D: Area = 22 cm^2 Perimeter = 24 cm
Yes, all the figures have the same area.
No, they do not have the same perimeter.

1. 86
   86 m
2. 140
   140 in.

Workbook

Exercise 4, pp. 172-173 (answers p. 162)

Enrichment

Ask your student if she can think of a different method for finding the perimeter of these figures. Discuss a short-cut method. We can imagine the figures are formed from a rectangle with part of the sides "pushed in." The perimeter of the figure would be the same as that of the rectangle, plus some extra sides if one side of the rectangle was "pushed in" in the middle away from the corners.

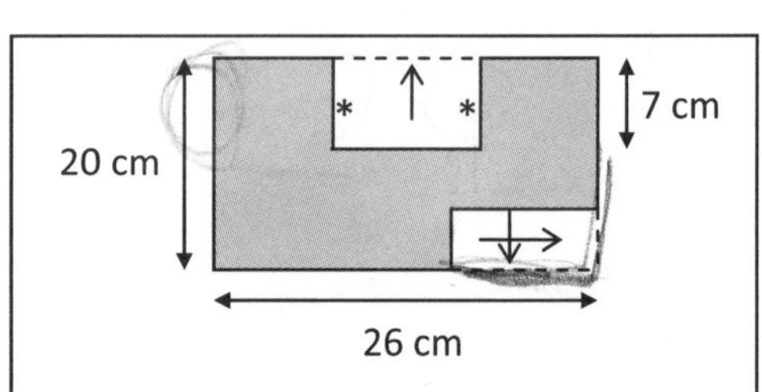

The perimeter of the figure is the same as that of a rectangle with sides 26 cm and 20 cm, plus 2 x 7 cm for the starred sides.

(2) Find the area of a composite figure

Discussion

Tasks 3-4, p. 153

After your student has found the answers to these, discuss some alternate methods.

3: The figure could also be broken up into two squares 5 m by 5 m and one rectangle in the middle 4 m by 10 m, as shown at the right.

Or, we could find the area of a large rectangle 14 m by 10 m, and then subtract the area of two smaller squares 5 m by 5 m.

4: The figure could also be divided up into three rectangles in two different ways, and the areas of the three rectangles added together.

3. $70\text{ m}^2 + 20\text{ m}^2 =$ **90** m^2

4. $120\text{ yd}^2 - 15\text{ yd}^2 =$ **105** yd^2

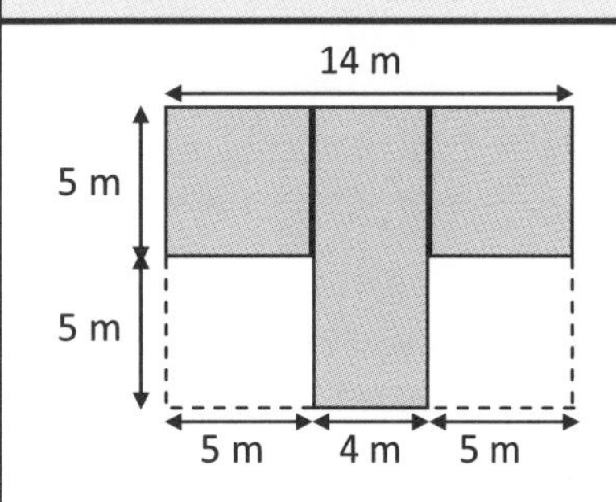

Area = $25\text{ m}^2 + 25\text{ m}^2 + 40\text{ m}^2 = 90\text{ m}^2$

Area = $140\text{ m}^2 - 25\text{ m}^2 - 25\text{ m}^2 = 90\text{ m}^2$

Workbook

Exercise 5, pp. 174 (answers p. 162)

Enrichment

Draw the figure at the right, which is made up of two squares and a circle. Tell your student that the area of the circle is 2463 cm^2 and ask him to find the area of the shaded part.

The area of the shaded parts around the outside of the circle is simply the difference between the area of the square and the area of the circle, which is given. The difficult part is finding the area of the inside square, since we don't know the length of the sides.

To find the area of the inside square, we can divide the figure into four smaller squares. Then each quarter of the inside square has half of the area of a square with sides of 28 cm.

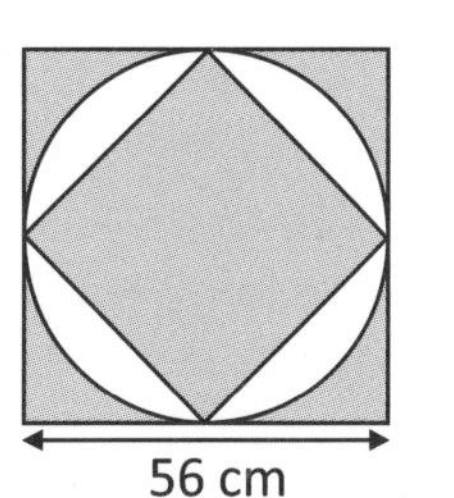

Area of circle $= 2463\text{ cm}^2$

Area of shaded part = ?

Area of all 4 corner pieces:
$(56\text{ cm} \times 56\text{ cm}) - 2463\text{ cm}^2 = 673\text{ cm}^2$

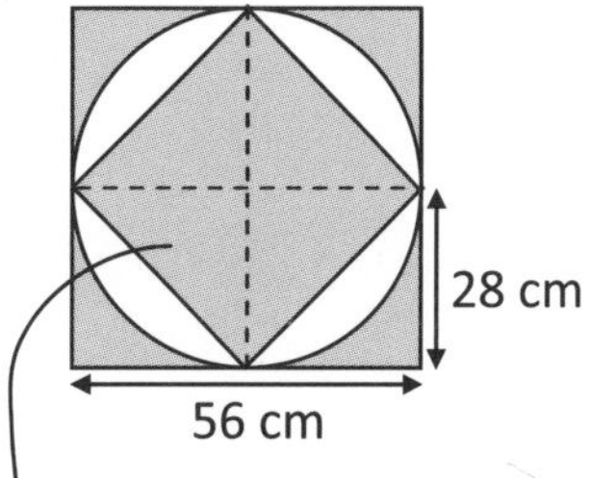

Area = $(28\text{ cm} \times 28\text{ cm}) \div 2 = 392\text{ cm}^2$

Total area of shaded part:
$673\text{ cm}^2 + (4 \times 392\text{ cm}^2) = 2241\text{ cm}^2$

(3) Find the area of paths

Activity

Cut out two rectangles from centimeter graph paper, one smaller than the other, for example 5 cm by 8 cm and 3 cm by 6 cm. Ask your student to find the difference in area of these two rectangles.

Place the smaller rectangle on top of the other and move it around, showing that the difference in the area of the two is always 22 cm^2.

Draw a rectangle with a path around it, giving the length of the larger rectangle, the width of the smaller rectangle, and the width of the "path." Tell your student that the width, or margin, of the path is the same all the way around.

Ask your student to find the area of the larger rectangle, the smaller rectangle, and then the path. Note that the width of the path needs to be taken into account on *both* sides.

Discuss other ways of finding this path. It can be found by dividing it up into smaller rectangles, such as two rectangles that are 10 cm by 2 cm and two that are 4 cm by 2 cm, and adding the areas together. Ask which method is easier.

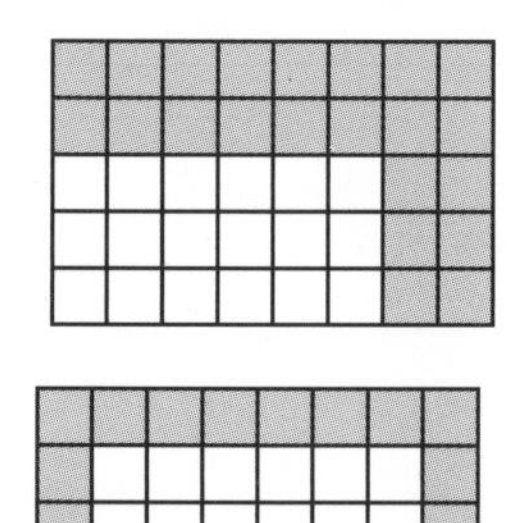

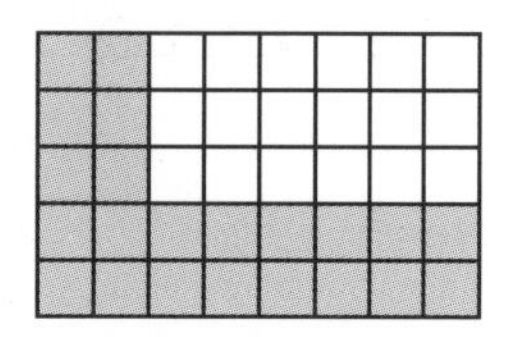

Area of larger rectangle
= 5 cm x 8 cm = 40 cm^2
Area of smaller rectangle
= 3 cm x 6 cm = 18 cm^2
Shaded area = Difference
= 40 cm^2 – 18 cm^2 = 22 cm^2

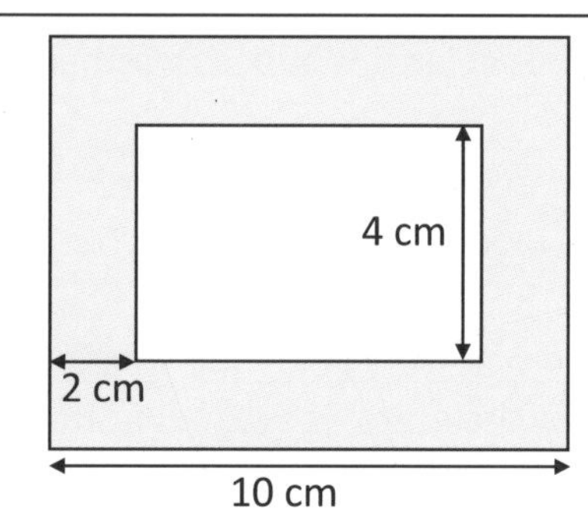

Width of large rectangle = 4 cm + (2 cm x 2) = 8 cm
Area of large rectangle = 10 cm x 8 cm = 80 cm^2
Length of small rectangle = 10 cm – (2 cm x 2) = 6 cm
Area of smaller rectangle = 6 cm x 4 cm = 24 cm^2
Area of "path" = 80 cm^2 – 24 cm^2 = 56 cm^2

Practice

Tasks 5-6, p. 154

Workbook

Exercise 6, pp. 175-176 (answers p. 162)

Reinforcement

Extra Practice, Unit 5, Exercise 3, pp. 91-94

5. Area of big rectangle = **80** m^2
   Area of small rectangle = 6 m x 8 m = **48** m^2
   Area of path = 80 m^2 – 48 m^2 = **32** m^2

6. Area of big rectangle = **120** yd^2
   Length of small rectangle = 12 yd – 2 yd – 3 yd = **7** yd
   Width of small rectangle = 10 yd – 2 yd – 2 yd = **6** yd
   Area of small rectangle = 7 yd x 6 yd = **42** yd^2
   Area of shaded part = 120 yd^2 – 42 yd^2 = **78** yd^2

(4) Practice

Practice

Practice C, pp. 155-156

Test

Tests, Unit 5, 3A and 3B, pp. 179-182

1. (a) Area: 40 ft^2
   Perimeter: 28 ft
   (b) Area: 59 m^2
   Perimeter: 40 m
   (c) Area: 168 m^2
   Perimeter: 68 m
   (d) Area: 483 km^2
   Perimeter: 112 km
2. (a) 294 m^2
   (b) 642 cm^2
   (c) 440 $in.^2$
   (d) 225 yd^2
3. (48 cm x 48 cm) – (40 cm x 40 cm) = **704 cm^2**
   The area of the border is 704 cm^2.
4. (14 m x 10 m) – (10 m x 6 m) = **80 m^2**
   The area of the path is 80 m^2.
5. (68 cm x 54 cm) – (60 cm x 46 cm) = **912 cm^2**
   The area of the table not covered is 912 cm^2.
9. 6 m x 5 m = **30 m^2**
   The area of carpet is 30 m^2.

Workbook

Exercise 4, pp. 172-173

1. (a) $9 + 9 + 9 + 6 + 15 + 6 + 3 + 9 =$ **66 cm**
   Or: $2 \times (18 + 15) = 66$ cm
   (b) $15 + 6 + 6 + 3 + 9 + 9 =$ **48 m**
   Or: $2 \times (15 + 9) = 48$ m
   (c) $8 + 6 + 10 + 5 + 10 + 6 +$
   $8 + 6 + 10 + 5 + 10 + 6 =$ **90 in.**
   Or: $2 \times (28 + 17) = 90$ in.

2. (a) $8 + 6 + 10 + 4 + 10 + 6 + 8 + 16 =$ **68 cm**
   Or: $2 \times (18 + 16) = 68$ cm
   (b) $5 + 5 + 10 + 10 + 25 + 5 + 10 + 10 =$ **80 m**
   Or: $2 \times (25 + 15) = 80$ m
   (Note for 2008 printing: arrow at right should be just for length of shaded part.)
   (c) $8 + 7 + 10 + 7 + 8 + 13 + 8 + 7 + 18 + 20 =$ **106 m**
   Or: $2 \times (26 + 20) + (2 \times 7) = 106$ m

Exercise 5, p. 174

Solutions will vary. Solutions which give the fewest rectangles are given here.

1. (a)

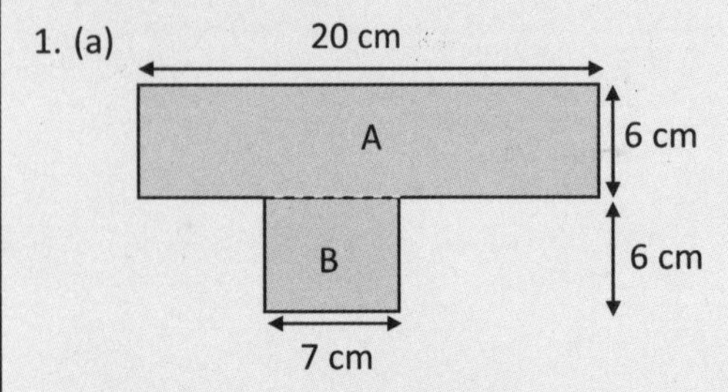

Area of A: 20 cm x 6 cm = 120 cm^2
Area of B: 7 cm x 6 cm = 42 cm^2
Total area: 120 cm^2 + 42 cm^2 = **162 cm^2**

(b)

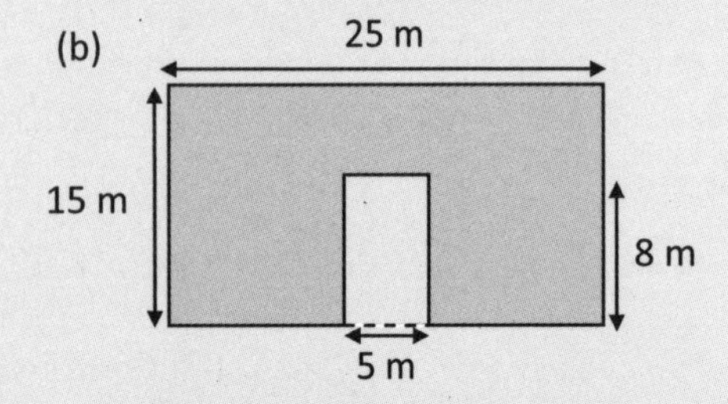

Area of large rectangle: 15 m x 25 m = 375 m^2
Area of small rectangle: 5 m x 8 m = 40 m^2
Area of shaded part: 375 m^2 – 40 m^2 = **335 m^2**

(c)

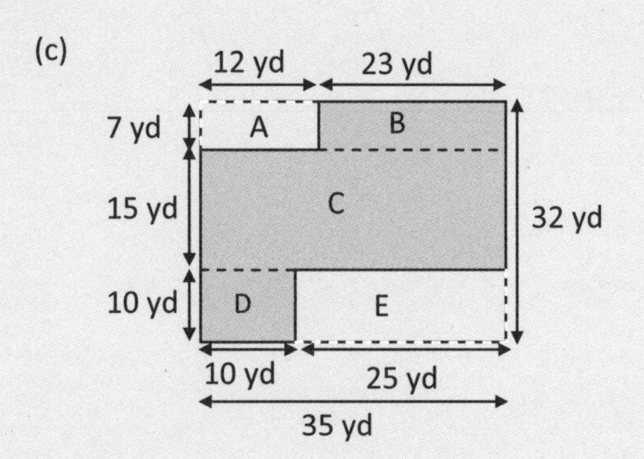

Area of large rectangle: 35 yd x 32 yd = 1120 yd^2
Area of A: 12 yd x 7 yd = 84 yd^2
Area of E: 25 yd x 10 yd = 250 yd^2
Area of shaded part:
1120 yd^2 – 84 yd^2 – 250 yd^2 = **786 yd^2**
or
Area of B: 7 yd x 23 yd = 161 yd^2
Area of C: 15 yd x 35 yd = 525 yd^2
Area of D: 10 yd x 10 yd = 100 yd^2
Area of shaded part:
161 yd^2 + 525 yd^2 + 100 yd^2 = 786 yd^2

Exercise 6, pp. 175-176

1. (a) (16 m x 8 m) – (12 m x 4 m)
   = 128 m^2 – 48 m^2 = **80 m^2**
   (b) (9 m x 5 m) – (3 m x 3 m)
   45 m^2 – 9 m^2 = **36 m^2**
   (d) (20 in. x 16 in.) – (8 in. x 5 in.)
   = 320 $in.^2$ – 40 $in.^2$ = **280 $in.^2$**

2. Area of pool + path: 22 m x 14 m = 308 m^2
   Area of pool: 20 m x 12 m = 240 m^2
   Area of path: 308 m^2 – 240 m^2 = **68 m^2**

3. Area of towel: 60 cm x 96 cm = 5760 cm^2
   Area without border: 54 cm x 90 cm = 4860 cm^2
   Area of border = 5760 cm^2 – 4860 cm^2 = **900 cm^2**

Review 5

Review

Review 5, pp. 157-161

Workbook

Review 5 pp. 177-184 (answers p. 164)

Test

Tests, Units 1-5, Cumulative A and B, pp. 183-194

1. (a) 741 (b) 1056 (c) 396
   (d) 6448 (e) 14,336 (f) 1188
2. 108 ÷ 6 = **18**; The other number is 18.
3. (a) Any 3: $\frac{4}{6}, \frac{6}{9}, \frac{8}{12}, \frac{10}{15}, \frac{12}{18}, \ldots$
   (b) Any 3: $\frac{2}{10}, \frac{3}{15}, \frac{4}{20}, \frac{5}{25}, \frac{6}{30}, \ldots$
   (c) Any 3: $\frac{3}{4}, \frac{6}{8}, \frac{12}{16}, \frac{15}{20}, \frac{18}{24}, \ldots$
4. (a) $\frac{2}{3}, \frac{3}{4}, \frac{5}{6}$ (b) $1\frac{7}{10}, \frac{7}{4}, 2$

5. (a) $\frac{3}{4}$ (b) $\frac{3}{4}$ (c) $\frac{2}{5}$
6. (a) > (b) = (c) >
7. 0, –5, –10
8. 7, 11, 13
9. –11, –6, –2, 0, 3, 6, 7
10. (a) 6 (b) 6 (c) 5 (d) 5 (e) 4 (f) 2
11. Always true
    Always true
    Always true
    Sometimes true
12. (a) 8014 – 4362 = **3652** (b) 2335 – 709 = **1626**
    (c) 620 – 309 = **311** (d) 499 + 377 = **876**
    (e) 2142 ÷ 6 = **357** (f) 3340 ÷ 4 = **835**
    (g) 7200 ÷ 8 = **900** (h) 715 x 7 = **5005**
13. 100¢ – (3 x 25¢) = 100¢ – 75¢ = 25¢
14. 348 x 39 = **13,572**
15. 

    5 units = $70
    1 unit = $70 ÷ 5 = $14
    8 units = $14 x 8 = **$112**
    She would save $112 in 8 weeks.
16. $\frac{9\text{ m}}{4} = \mathbf{2\frac{1}{4}\text{ m}}$
    Each part is $2\frac{1}{4}$ m long.
17. Triangle OSR is an isosceles triangle.

18. A, D
19. (a) A, C
    (b) B, C
    (c) C
20. H, Z
    Z
21. Length = 78 m^2 ÷ 6 m = **13 m**
    Perimeter = 13 m + 13 m + 6 m + 6 m = **38 m**
22. Side = 48 cm ÷ 4 = 12 cm
    Area = 12 cm x 12 cm = **144 cm^2**
23. (a) 156 m^2 (b) 108 m^2
24. (a) Area = 460 cm^2 Perimeter = 110 cm
    (b) Area = 128 m^2 Perimeter = 60 m
25. $4\frac{2}{5}$ yd – $\frac{3}{10}$ yd = $4\frac{4}{10}$ yd – $\frac{3}{10}$ yd = $4\frac{1}{10}$ yd
    = $4\frac{2}{5}$ yd + $4\frac{1}{10}$ yd = $4\frac{4}{10}$ yd + $4\frac{1}{10}$ yd = **$8\frac{1}{2}$ yd**
    Together they used $8\frac{1}{2}$ yd of rope.
26. 1250 in. – (75 in. x 12) = 1250 in. – 900 in. = **350 in.**
    The remaining piece is 350 in. long.
27. 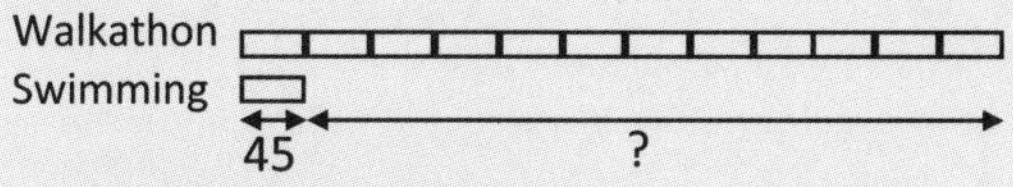

    11 units = 45 x 11 = **495**
    495 more people took part in the walkathon.
18. Amount sold = (138 x 24) – 72 = 3240
    3240 ÷ 3 = **1080**
    Each bought 1080 mangoes.

Workbook

Review 5, pp. 177-184

1. (a) \$5703
   (b) \$34,000,864
   (c) –140

2. hundred

3. 100

4. 1,534,502

5. (a) –3
   (b) –8
   (c) –1
   (d) 999,998
   (e) 199,999,800

6. (a) <
   (b) <
   (c) <
   (d) > (264 – 2000 will be negative)
   (e) < (37 – 15) x (8 + 14) = 22 x 22
   (f) >

7. (a) 6000
   (b) 4 (24 = 6 x 4)
   (c) 9 (2 x 18 = 2 x 2 x 9 = 4 x 9)
   (d) 16 (16 + 4 x 8 = 16 + 32)
   (e) 7 (12 ÷ 4 + 9 – 2 = 3 + 9 – 2 = 3 + 7)

8. $\frac{23}{5}$

9. 10 out of 50 = $\frac{10}{50}$ = $\mathbf{\frac{1}{5}}$

   $\frac{1}{5}$ of her coins are dimes. (Error 2008 printing, *money* should be *coins*.)

10. $\frac{3}{4}$ ℓ or 750 ml

11. 24 cm

12. 14 cm x 3 = **42 cm**
    The perimeter is 42 cm.

13. 2 x (10 in. + 4 in.) = **28 in.**
    The parallelogram has a perimeter of 28 in.

14. Area of each square is 4 cm^2.
    There are 16 squares.
    Total area: 4 cm^2 x 16 = **64 cm^2**

15. (a) Mary
    (b) Dani

16. (a) $\angle x$ = 145°
    $\angle y$ = 35°
    (b) $\angle x$ = 308°
    $\angle y$ = 52°

17. (a) PQ
    (b) CD

18. (a) 160 cm (b) 28 yd

19. (a) False
    (b) True
    (c) False
    (d) False
    (e) True

20. 500

21. Equilateral triangle

22. (a) Thursday
    (b) 630

23. 8 cm

24. Perimeter = 28 m
    Area = 32 m^2

25. C

26. } 1080 in.

    3 units = 1080 in.
    1 unit = 1080 in. ÷ 3 = **360 in.**
    The length of the shorter piece is 360 in.

27. 216 m

28. Area of rectangle: 9 cm x 4 cm = 36 cm^2
    Area of square: 36 cm^2
    6 x 6 = 36.
    The length of side of square = **6 cm**

17. Area of room: 8 m x 6 m = 48 m^2
    Area of carpet: 6 m x 4 m = 24 m^2
    48 m^2 – 24 m^2 = **24 m^2**
    The area not covered is 24 m^2.

Mental Math 1	Mental Math 2	Mental Math 3		Mental Math 4	Mental Math 5
37 + 6 = **43**	41 + 90 = **131**	92 – 9 = **83**		98 – 39 = **59**	1400 + 300 = **1700**
48 + 7 = **55**	78 + 48 = **126**	84 – 5 = **79**		84 – 55 = **29**	17,000 + 9000 = **26,000**
69 + 9 = **78**	49 + 28 = **77**	73 – 9 = **64**		73 – 29 = **44**	890,000 + 30,000 = **920,000**
46 + 5 = **51**	48 + 37 = **85**	50 – 8 = **42**		50 – 18 = **32**	4,600,000 + 500,000 = **5,100,000**
32 + 5 = **37**	22 + 53 = **75**	93 – 7 = **86**		93 – 72 = **21**	320,000 + 80,000 = **400,000**
16 + 8 = **24**	38 + 81 = **119**	66 – 9 = **57**		66 – 49 = **17**	4,500,000 + 2,600,000 = **7,100,000**
92 + 8 = **100**	79 + 34 = **113**	53 – 5 = **48**		53 – 35 = **18**	23,000 + 60,000 = **83,000**
34 + 6 = **40**	39 + 27 = **66**	60 – 6 = **54**		60 – 46 = **14**	4,500,000 + 5,000,000 = **9,500,000**
98 + 3 = **101**	16 + 45 = **61**	72 – 8 = **64**		72 – 38 = **34**	28,000 + 63,000 = **91,000**
65 + 7 = **72**	25 + 49 = **74**	55 – 9 = **46**		55 – 29 = **26**	320,000 + 430,000 = **750,000**
62 + 9 = **71**	48 + 55 = **103**	60 – 4 = **56**		90 – 44 = **46**	456,000 + 300,000 = **756,000**
81 + 6 = **87**	89 + 39 = **128**	79 – 3 = **76**		92 – 37 = **55**	240,000 + 60,000 = **300,000**
25 + 8 = **33**	26 + 74 = **100**	22 – 5 = **17**		22 – 15 = **7**	45,800 + 900 = **46,700**
91 + 7 = **98**	27 + 33 = **60**	88 – 6 = **82**		88 – 56 = **32**	23,000 + 7000 = **30,000**
34 + 3 = **37**	11 + 27 = **38**	32 – 7 = **25**		32 – 17 = **15**	920,000 + 500,000 = **1,420,000**
439 + 3 = **442**	84 + 63 = **147**	682 – 57 = **625**		77 – 39 = **38**	45,300 + 150 = **45,450**
552 + 8 = **560**	68 + 76 = **144**	195 – 68 = **127**		67 – 18 = **49**	270,000 + 710,000 = **980,000**
619 + 4 = **623**	77 + 47 = **124**	277 – 39 = **238**		86 – 28 = **58**	450,000 + 32,000 = **482,000**
778 + 6 = **784**	58 + 69 = **127**	967 – 18 = **949**		60 – 15 = **45**	400,001 + 99 = **400,100**
255 + 8 = **263**	56 + 79 = **135**	486 – 28 = **458**		100 – 37 = **63**	809,001 + 999 = **810,000**

Mental Math 6	Mental Math 7		Mental Math 8	Mental Math 9
51,000 – 9000 = **42,000**	8000 x 8 = **64,000**		24,000 ÷ 3 = **8000**	4,000,000 x 3 = **12,000,000**
420,000 – 60,000 = **360,000**	800,000 x 3 = **2,400,000**		49,000,000 ÷ 7 = **7,000,000**	49,000 – 7000 = **42,000**
3,800,000 – 2,600,000 = **1,200,000**	80,000 x 9 = **720,000**		270,000 ÷ 9 = **30,000**	2700 + 9000 = **11,700**
840,000 – 170,000 = **670,000**	5,000,000 x 5 = **25,000,000**		32,000 ÷ 8 = **4000**	40,000 ÷ 8 = **5000**
82,000 – 5000 = **77,000**	6000 x 9 = **54,000**		1,800,000 ÷ 2 = **900,000**	180,000 ÷ 10 = **18,000**
49,000 – 6000 = **43,000**	5 x 60,000 = **300,000**		200,000 ÷ 5 = **40,000**	4000 x 4 = **16,000**
10,000 – 100 = **9,900**	500,000 x 7 = **3,500,000**		54,000 ÷ 9 = **6000**	3,200,000 – 500,000 = **2,700,000**
2,300,000 – 900,000 = **1,400,000**	7000 x 8 = **56,000**		1,800,000 ÷ 6 = **300,000**	54,000 + 92,000 = **146,000**
820,000 – 30,000 = **790,000**	7 x 700 = **4900**		40,000 ÷ 8 = **5000**	9800 – 3600 = **6200**
51,000 – 6000 = **45,000**	800,000 x 6 = **4,800,000**		3500 ÷ 7 = **500**	4,200,000 ÷ 7 = **600,000**
920,000 – 460,000 = **460,000**	9 x 9000 = **81,000**		36,000 ÷ 4 = **9000**	4 x 900 = **3600**
54,000 – 9,000 = **45,000**	60,000 x 6 = **360,000**		300,000 ÷ 6 = **50,000**	5,600,000 ÷ 7 = **800,000**
59,000 – 37,000 = **22,000**	20,000 x 3 = **60,000**		2,100,000 ÷ 7 = **300,000**	3000 ÷ 6 = **500**
13,400 – 800 = **12,600**	40,000 x 8 = **320,000**		15,000 ÷ 5 = **3000**	260,000 – 7000 = **253,000**
62,300 – 7000 = **55,300**	20,000 x 4 = **80,000**		160,000 ÷ 4 = **40,000**	15,000 – 50 = **14,950**
892,000 – 7000 = **885,000**	3000 x 7 = **21,000**		8,100,000 ÷ 9 = **900,000**	160,000 + 2000 = **162,000**
2,400,000 – 50,000 = **2,350,000**	3 x 30,000 = **90,000**		4800 ÷ 8 = **600**	33,400 + 800 = **34,200**
24,100 – 800 = **23,300**	300,000 x 9 = **2,700,000**		24,000 ÷ 6 = **4000**	910,000 – 300,000 = **610,000**
693,000 – 6000 = **687,000**	2 x 50,000 = **100,000**		450,000 ÷ 9 = **50,000**	2 x 200,000 = **400,000**
35,100 – 9000 = **26,100**	300 x 4 = **1200**		36,000 ÷ 6 = **6000**	91,000 ÷ 1000 = **91**

Mental Math 10	Mental Math 11	Mental Math 12	Mental Math 13	Mental Math 14	Mental Math 15
28 + 7 = **35**	1234 + 9 = **1243**	8943 + 99 = **9042**	45 x 9 = **405**	99 x 42 = **4158**	90 x 80 = **7200**
4128 + 7 = **4135**	1234 + 99 = **1333**	5861 – 39 = **5822**	45 x 99 = **4455**	199 x 42 = **8358**	500 x 60 = **30,000**
4281 + 70 = **4351**	1234 + 999 = **2233**	892 + 198 = **1090**	45 x 999 = **44,955**	25 x 199 = **4975**	900 x 600 = **540,000**
2841 + 700 = **3541**	1234 + 499 = **1733**	575 + 29 = **604**	45 x 20 = **900**	14 x 35 = **490**	2000 x 50 = **100,000**
2841 + 704 = **3545**	1234 – 9 = **1225**	7003 – 997 = **6006**	45 x 19 = **855**	45 x 12 = **540**	700 x 70 = **49,000**
8451 – 7 = **8444**	1234 – 99 = **1135**	4863 – 99 = **4764**	32 x 25 = **800**	22 x 35 = **770**	800 x 80 = **64,000**
8514 – 70 = **8444**	1234 – 999 = **235**	3000 – 345 = **2655**	320 x 25 = **8000**	101 x 89 = **8989**	80 x 6 = **480**
5148 – 700 = **4448**	1234 – 199 = **1035**	4569 + 2200 = **6769**	99 x 8 = **792**	202 x 12 = **2424**	30 x 40 = **1200**
5148 – 704 = **4444**	4028 + 8 = **4036**	8107 – 4900 = **3207**	99 x 98 = **9702**	15 x 18 = **270**	60 x 7000 = **420,000**
4517 + 59 = **4576**	4028 + 98 = **4126**	4900 – 63 = **4837**	4 x 98 = **392**	25 x 28 = **700**	40 x 60 = **2400**
6784 – 59 = **6725**	4028 + 998 = **5026**	1928 + 5900 = **7828**	24 x 25 = **600**	46 x 99 = **4554**	2000 x 3000 = **6,000,000**
5418 + 48 = **5466**	8761 – 8 = **8753**	3872 + 990 = **4862**	894 x 9 = **8046**	46 x 25 = **1150**	700 x 50 = **35,000**
1862 – 48 = **1814**	8761 – 98 = **8663**	3872 + 996 = **4868**	25 x 64 = **1600**	460 x 250 = **115,000**	3000 x 80 = **240,000**
300 – 42 = **258**	8761 – 998 = **7763**	290 + 998 = **1288**	30 x 500 = **15,000**	11 x 10 = **110**	800 x 400 = **320,000**
2300 – 42 = **2258**	3985 + 7 = **3992**	642 – 58 = **584**	998 x 5 = **4990**	11 x 11 = **121**	90 x 400 = **36,000**
3000 – 420 = **2580**	3985 + 97 = **4082**	777 + 999 = **1776**	2060 x 8 = **16,480**	11 x 12 = **132**	30 x 900 = **27,000**
3008 – 420 = **2588**	3985 + 997 = **4982**	2222 – 999 = **1223**	8 x 59 = **472**	22 x 12 = **264**	90 x 70 = **6300**
8600 – 81 = **8519**	7643 – 7 = **7636**	2905 + 996 = **3901**	48 x 25 = **1200**	12 x 75 = **900**	50 x 400 = **20,000**
9000 – 350 = **8650**	7643 – 97 = **7546**	6000 – 556 = **5444**	99 x 57 = **5643**	999 x 55 = **54,945**	90 x 5000 = **450,000**
9000 – 1234 = **7766**	7643 – 997 = **6646**	2080 + 5188 = **7268**	999 x 99 = **98,901**	1999 x 55 = **109,945**	200 x 900 = **180,000**

Mental Math 16	Mental Math 17	Mental Math 18	Mental Math 19	Mental Math 20	Mental Math 21
$\frac{1}{2}+\frac{1}{4}=\mathbf{\frac{3}{4}}$	$\frac{1}{3}-\frac{1}{6}=\mathbf{\frac{1}{6}}$	$\frac{5}{12}+\frac{1}{6}=\mathbf{\frac{7}{12}}$	$\frac{7}{10}+\frac{2}{5}=\mathbf{1\frac{1}{10}}$	$5-1\frac{1}{8}=\mathbf{3\frac{7}{8}}$	$\frac{1}{5}\times 35=\mathbf{7}$
$\frac{1}{8}+\frac{1}{2}=\mathbf{\frac{5}{8}}$	$\frac{1}{2}-\frac{1}{4}=\mathbf{\frac{1}{4}}$	$\frac{1}{15}+\frac{2}{5}=\mathbf{\frac{7}{15}}$	$2-\frac{7}{10}=\mathbf{1\frac{3}{10}}$	$4-2\frac{1}{5}=\mathbf{1\frac{4}{5}}$	$\frac{1}{2}\times 24=\mathbf{12}$
$\frac{1}{3}+\frac{1}{12}=\mathbf{\frac{5}{12}}$	$\frac{1}{2}-\frac{3}{8}=\mathbf{\frac{1}{8}}$	$\frac{1}{10}+\frac{3}{100}=\mathbf{\frac{13}{100}}$	$\frac{5}{12}+\frac{5}{6}=\mathbf{1\frac{1}{4}}$	$3-1\frac{1}{4}=\mathbf{1\frac{3}{4}}$	$\frac{1}{8}\times 16=\mathbf{2}$
$\frac{1}{4}+\frac{1}{8}=\mathbf{\frac{3}{8}}$	$\frac{1}{2}-\frac{1}{6}=\mathbf{\frac{1}{3}}$	$\frac{3}{10}-\frac{1}{100}=\mathbf{\frac{29}{100}}$	$7-\frac{3}{4}=\mathbf{6\frac{1}{4}}$	$6-3\frac{1}{3}=\mathbf{2\frac{2}{3}}$	$\frac{1}{4}\times 40=\mathbf{10}$
$\frac{1}{12}+\frac{1}{6}=\mathbf{\frac{1}{4}}$	$\frac{2}{3}-\frac{5}{9}=\mathbf{\frac{1}{9}}$	$\frac{7}{10}-\frac{3}{20}=\mathbf{\frac{11}{20}}$	$\frac{7}{12}+\frac{3}{4}=\mathbf{1\frac{1}{3}}$	$10-1\frac{3}{4}=\mathbf{8\frac{1}{4}}$	$\frac{7}{10}\times 30=\mathbf{21}$
$\frac{1}{3}+\frac{1}{6}=\mathbf{\frac{1}{2}}$	$\frac{2}{3}-\frac{1}{12}=\mathbf{\frac{7}{12}}$	$\frac{1}{3}-\frac{5}{21}=\mathbf{\frac{2}{21}}$	$\frac{7}{9}+\frac{5}{18}=\mathbf{1\frac{1}{18}}$	$7-2\frac{2}{5}=\mathbf{4\frac{3}{5}}$	$\frac{2}{3}\times 18=\mathbf{12}$
$\frac{1}{2}+\frac{1}{10}=\mathbf{\frac{3}{5}}$	$\frac{4}{5}-\frac{7}{10}=\mathbf{\frac{1}{10}}$	$\frac{16}{35}+\frac{3}{7}=\mathbf{\frac{31}{35}}$	$3-\frac{7}{12}=\mathbf{2\frac{5}{12}}$	$5-4\frac{3}{5}=\mathbf{\frac{2}{5}}$	$\frac{3}{4}\times 32=\mathbf{24}$
$\frac{1}{9}+\frac{1}{3}=\mathbf{\frac{4}{9}}$	$\frac{3}{4}-\frac{3}{8}=\mathbf{\frac{3}{8}}$	$\frac{2}{3}+\frac{3}{18}=\mathbf{\frac{5}{6}}$	$6-\frac{2}{5}=\mathbf{5\frac{3}{5}}$	$10-3\frac{1}{10}=\mathbf{6\frac{9}{10}}$	$\frac{4}{5}\times 25=\mathbf{20}$
$\frac{1}{2}+\frac{1}{6}=\mathbf{\frac{2}{3}}$	$\frac{5}{6}-\frac{5}{12}=\mathbf{\frac{5}{12}}$	$\frac{5}{12}-\frac{7}{60}=\mathbf{\frac{3}{10}}$	$\frac{9}{11}+\frac{9}{22}=\mathbf{1\frac{5}{22}}$	$12-11\frac{4}{9}=\mathbf{\frac{5}{9}}$	$\frac{5}{7}\times 14=\mathbf{10}$
$\frac{1}{12}+\frac{1}{2}=\mathbf{\frac{7}{12}}$	$\frac{1}{2}-\frac{3}{10}=\mathbf{\frac{1}{5}}$	$\frac{1}{24}+\frac{1}{8}=\mathbf{\frac{1}{6}}$	$5-\frac{7}{8}=\mathbf{4\frac{1}{8}}$	$15-10\frac{89}{100}=\mathbf{4\frac{11}{100}}$	$\frac{3}{11}\times 66=\mathbf{18}$
$\frac{1}{12}+\frac{1}{4}=\mathbf{\frac{1}{3}}$	$\frac{7}{8}-\frac{3}{4}=\mathbf{\frac{1}{8}}$	$\frac{2}{9}+\frac{5}{18}=\mathbf{\frac{1}{2}}$	$\frac{2}{3}+\frac{11}{12}=\mathbf{1\frac{7}{12}}$	$\frac{1}{6}+\mathbf{3\frac{5}{6}}=4$	$\frac{3}{8}\times 56=\mathbf{21}$
$\frac{1}{10}+\frac{1}{5}=\mathbf{\frac{3}{10}}$	$\frac{7}{12}-\frac{1}{2}=\mathbf{\frac{1}{12}}$	$\frac{47}{50}-\frac{2}{5}=\mathbf{\frac{27}{50}}$	$\frac{13}{24}+\frac{5}{8}=\mathbf{1\frac{1}{6}}$	$2\frac{1}{6}+\mathbf{1\frac{5}{6}}=4$	$\frac{2}{3}\times 30=\mathbf{20}$
$\frac{2}{3}+\frac{1}{9}=\mathbf{\frac{7}{9}}$	$\frac{11}{12}-\frac{3}{4}=\mathbf{\frac{1}{6}}$	$\frac{5}{6}-\frac{5}{18}=\mathbf{\frac{5}{9}}$	$\frac{5}{7}+\frac{17}{21}=\mathbf{1\frac{11}{21}}$	$7\frac{4}{5}+\mathbf{2\frac{1}{5}}=10$	$\frac{5}{12}\times 24=\mathbf{10}$
$\frac{1}{4}+\frac{3}{8}=\mathbf{\frac{5}{8}}$	$\frac{7}{10}-\frac{2}{5}=\mathbf{\frac{3}{10}}$	$\frac{3}{4}+\frac{3}{24}=\mathbf{\frac{7}{8}}$	$4-\frac{5}{18}=\mathbf{3\frac{13}{18}}$	$3\frac{4}{9}+\mathbf{4\frac{5}{9}}=8$	$\frac{3}{5}\times 100=\mathbf{60}$
$\frac{1}{5}+\frac{7}{10}=\mathbf{\frac{9}{10}}$	$\frac{2}{3}-\frac{2}{9}=\mathbf{\frac{4}{9}}$	$\frac{4}{5}-\frac{9}{40}=\mathbf{\frac{23}{40}}$	$12-\frac{7}{9}=\mathbf{11\frac{2}{9}}$	$\mathbf{1\frac{1}{6}}+5\frac{5}{6}=7$	$\frac{3}{4}\times 100=\mathbf{75}$

Mental Math 1	Mental Math 2	Mental Math 3
37 + 6 = ________	41 + 90 = ________	92 – 9 = ________
48 + 7 = ________	78 + 48 = ________	84 – 5 = ________
69 + 9 = ________	49 + 28 = ________	73 – 9 = ________
46 + 5 = ________	48 + 37 = ________	50 – 8 = ________
32 + 5 = ________	22 + 53 = ________	93 – 7 = ________
16 + 8 = ________	38 + 81 = ________	66 – 9 = ________
92 + 8 = ________	79 + 34 = ________	53 – 5 = ________
34 + 6 = ________	39 + 27 = ________	60 – 6 = ________
98 + 3 = ________	16 + 45 = ________	72 – 8 = ________
65 + 7 = ________	25 + 49 = ________	55 – 9 = ________
62 + 9 = ________	48 + 55 = ________	60 – 4 = ________
81 + 6 = ________	89 + 39 = ________	79 – 3 = ________
25 + 8 = ________	26 + 74 = ________	22 – 5 = ________
91 + 7 = ________	27 + 33 = ________	88 – 6 = ________
34 + 3 = ________	11 + 27 = ________	32 – 7 = ________
439 + 3 = ________	84 + 63 = ________	682 – 57 = ________
552 + 8 = ________	68 + 76 = ________	195 – 68 = ________
619 + 4 = ________	77 + 47 = ________	277 – 39 = ________
778 + 6 = ________	58 + 69 = ________	967 – 18 = ________
255 + 8 = ________	56 + 79 = ________	486 – 28 = ________

Mental Math 4	Mental Math 5
98 – 39 = ________	1400 + 300 = ____________
84 – 55 = ________	17,000 + 9000 = ____________
73 – 29 = ________	890,000 + 30,000 = ____________
50 – 18 = ________	4,600,000 + 500,000 = ____________
93 – 72 = ________	320,000 + 80,000 = ____________
66 – 49 = ________	4,500,000 + 2,600,000 = ___________
53 – 35 = ________	23,000 + 60,000 = ____________
60 – 46 = ________	4,500,000 + 5,000,000 = ___________
72 – 38 = ________	28,000 + 63,000 = ____________
55 – 29 = ________	320,000 + 430,000 = ____________
90 – 44 = ________	456,000 + 300,000 = ____________
92 – 37 = ________	240,000 + 60,000 = ____________
22 – 15 = ________	45,800 + 900 = ____________
88 – 56 = ________	23,000 + 7000 = ____________
32 – 17 = ________	920,000 + 500,000 = ____________
77 – 39 = ________	45,300 + 150 = ____________
67 – 18 = ________	270,000 + 710,000 = ____________
86 – 28 = ________	450,000 + 32,000 = ____________
60 – 15 = ________	400,001 + 99 = ____________
100 – 37 = ________	809,001 + 999 = ____________

Mental Math 6	Mental Math 7
51,000 – 9000 = ________	8000 x 8 = ________
420,000 – 60,000 = ________	800,000 x 3 = ________
3,800,000 – 2,600,000 = ________	80,000 x 9 = ________
840,000 – 170,000 = ________	5,000,000 x 5 = ________
82,000 – 5000 = ________	6000 x 9 = ________
49,000 – 6000 = ________	5 x 60,000 = ________
10,000 – 100 = ________	500,000 x 7 = ________
2,300,000 – 900,000 = ________	7000 x 8 = ________
820,000 – 30,000 = ________	7 x 700 = ________
51,000 – 6000 = ________	800,000 x 6 = ________
920,000 – 460,000 = ________	9 x 9000 = ________
54,000 – 9,000 = ________	60,000 x 6 = ________
59,000 – 37,000 = ________	20,000 x 3 = ________
13,400 – 800 = ________	40,000 x 8 = ________
62,300 – 7000 = ________	20,000 x 4 = ________
892,000 – 7000 = ________	3000 x 7 = ________
2,400,000 – 50,000 = ________	3 x 30,000 = ________
24,100 – 800 = ________	300,000 x 9 = ________
693,000 – 6000 = ________	2 x 50,000 = ________
35,100 – 9000 = ________	300 x 4 = ________

Mental Math 8	Mental Math 9
24,000 ÷ 3 = ____________	4,000,000 x 3 = ____________
49,000,000 ÷ 7 = ____________	49,000 – 7000 = ____________
270,000 ÷ 9 = ____________	2700 + 9000 = ____________
32,000 ÷ 8 = ____________	40,000 ÷ 8 = ____________
1,800,000 ÷ 2 = ____________	180,000 ÷ 10 = ____________
200,000 ÷ 5 = ____________	4000 x 4 = ____________
54,000 ÷ 9 = ____________	3,200,000 – 500,000 = __________
1,800,000 ÷ 6 = ____________	54,000 + 92,000 = ____________
40,000 ÷ 8 = ____________	9800 – 3600 = ____________
3500 ÷ 7 = ____________	4,200,000 ÷ 7 = ____________
36,000 ÷ 4 = ____________	4 x 900 = ____________
300,000 ÷ 6 = ____________	5,600,000 ÷ 7 = ____________
2,100,000 ÷ 7 = ____________	3000 ÷ 6 = ____________
15,000 ÷ 5 = ____________	260,000 – 7000 = ____________
160,000 ÷ 4 = ____________	15,000 – 50 = ____________
8,100,000 ÷ 9 = ____________	160,000 + 2000 = ____________
4800 ÷ 8 = ____________	33,400 + 800 = ____________
24,000 ÷ 6 = ____________	910,000 – 300,000 = ___________
450,000 ÷ 9 = ____________	2 x 200,000 = ____________
36,000 ÷ 6 = ____________	91,000 ÷ 1000 = ____________

Mental Math 10	Mental Math 11	Mental Math 12
28 + 7 = _______	1234 + 9 = _______	8943 + 99 = _______
4128 + 7 = _______	1234 + 99 = _______	5861 – 39 = _______
4281 + 70 = _______	1234 + 999 = _______	892 + 198 = _______
2841 + 700 = _______	1234 + 499 = _______	575 + 29 = _______
2841 + 704 = _______	1234 – 9 = _______	7003 – 997 = _______
8451 – 7 = _______	1234 – 99 = _______	4863 – 99 = _______
8514 – 70 = _______	1234 – 999 = _______	3000 – 345 = _______
5148 – 700 = _______	1234 – 199 = _______	4569 + 2200 = _______
5148 – 704 = _______	4028 + 8 = _______	8107 – 4900 = _______
4517 + 59 = _______	4028 + 98 = _______	4900 – 63 = _______
6784 – 59 = _______	4028 + 998 = _______	1928 + 5900 = _______
5418 + 48 = _______	8761 – 8 = _______	3872 + 990 = _______
1862 – 48 = _______	8761 – 98 = _______	3872 + 996 = _______
300 – 42 = _______	8761 – 998 = _______	290 + 998 = _______
2300 – 42 = _______	3985 + 7 = _______	642 – 58 = _______
3000 – 420 = _______	3985 + 97 = _______	777 + 999 = _______
3008 – 420 = _______	3985 + 997 = _______	2222 – 999 = _______
8600 – 81 = _______	7643 – 7 = _______	2905 + 996 = _______
9000 – 350 = _______	7643 – 97 = _______	6000 – 556 = _______
9000 – 1234 = _______	7643 – 997 = _______	2080 + 5188 = _______

Mental Math 13	Mental Math 14	Mental Math 15
45 x 9 = _______	99 x 42 = _______	90 x 80 = _______
45 x 99 = _______	199 x 42 = _______	500 x 60 = _______
45 x 999 = _______	25 x 199 = _______	900 x 600 = _______
45 x 20 = _______	14 x 35 = _______	2000 x 50 = _______
45 x 19 = _______	45 x 12 = _______	700 x 70 = _______
32 x 25 = _______	22 x 35 = _______	800 x 80 = _______
320 x 25 = _______	101 x 89 = _______	80 x 6 = _______
99 x 8 = _______	202 x 12 = _______	30 x 40 = _______
99 x 98 = _______	15 x 18 = _______	60 x 7000 = _______
4 x 98 = _______	25 x 28 = _______	40 x 60 = _______
24 x 25 = _______	46 x 99 = _______	2000 x 3000 = ______
894 x 9 = _______	46 x 25 = _______	700 x 50 = _______
25 x 64 = _______	460 x 250 = _______	3000 x 80 = _______
30 x 500 = _______	11 x 10 = _______	800 x 400 = _______
998 x 5 = _______	11 x 11 = _______	90 x 400 = _______
2060 x 8 = _______	11 x 12 = _______	30 x 900 = _______
8 x 59 = _______	22 x 12 = _______	90 x 70 = _______
48 x 25 = _______	12 x 75 = _______	50 x 400 = _______
99 x 57 = _______	999 x 55 = _______	90 x 5000 = _______
999 x 99 = _______	1999 x 55 = _______	200 x 900 = _______

Mental Math 16	Mental Math 17	Mental Math 18
$\frac{1}{2} + \frac{1}{4} =$	$\frac{1}{3} - \frac{1}{6} =$	$\frac{5}{12} + \frac{1}{6} =$
$\frac{1}{8} + \frac{1}{2} =$	$\frac{1}{2} - \frac{1}{4} =$	$\frac{1}{15} + \frac{2}{5} =$
$\frac{1}{3} + \frac{1}{12} =$	$\frac{1}{2} - \frac{3}{8} =$	$\frac{1}{10} + \frac{3}{100} =$
$\frac{1}{4} + \frac{1}{8} =$	$\frac{1}{2} - \frac{1}{6} =$	$\frac{3}{10} - \frac{1}{100} =$
$\frac{1}{12} + \frac{1}{6} =$	$\frac{2}{3} - \frac{5}{9} =$	$\frac{7}{10} - \frac{3}{20} =$
$\frac{1}{3} + \frac{1}{6} =$	$\frac{2}{3} - \frac{1}{12} =$	$\frac{1}{3} - \frac{5}{21} =$
$\frac{1}{2} + \frac{1}{10} =$	$\frac{4}{5} - \frac{7}{10} =$	$\frac{16}{35} + \frac{3}{7} =$
$\frac{1}{9} + \frac{1}{3} =$	$\frac{3}{4} - \frac{3}{8} =$	$\frac{2}{3} + \frac{3}{18} =$
$\frac{1}{2} + \frac{1}{6} =$	$\frac{5}{6} - \frac{5}{12} =$	$\frac{5}{12} - \frac{7}{60} =$
$\frac{1}{12} + \frac{1}{2} =$	$\frac{1}{2} - \frac{3}{10} =$	$\frac{1}{24} + \frac{1}{8} =$
$\frac{1}{12} + \frac{1}{4} =$	$\frac{7}{8} - \frac{3}{4} =$	$\frac{2}{9} + \frac{5}{18} =$
$\frac{1}{10} + \frac{1}{5} =$	$\frac{7}{12} - \frac{1}{2} =$	$\frac{47}{50} - \frac{2}{5} =$
$\frac{2}{3} + \frac{1}{9} =$	$\frac{11}{12} - \frac{3}{4} =$	$\frac{5}{6} - \frac{5}{18} =$
$\frac{1}{4} + \frac{3}{8} =$	$\frac{7}{10} - \frac{2}{5} =$	$\frac{3}{4} + \frac{3}{24} =$
$\frac{1}{5} + \frac{7}{10} =$	$\frac{2}{3} - \frac{2}{9} =$	$\frac{4}{5} - \frac{9}{40} =$

Mental Math 19	Mental Math 20	Mental Math 21
$\frac{7}{10} + \frac{2}{5} =$	$5 - 1\frac{1}{8} =$	$\frac{1}{5}$ x 35 =
$2 - \frac{7}{10} =$	$4 - 2\frac{1}{5} =$	$\frac{1}{2}$ x 24 =
$\frac{5}{12} + \frac{5}{6} =$	$3 - 1\frac{1}{4} =$	$\frac{1}{8}$ x 16 =
$7 - \frac{3}{4} =$	$6 - 3\frac{1}{3} =$	$\frac{1}{4}$ x 40 =
$\frac{7}{12} + \frac{3}{4} =$	$10 - 1\frac{3}{4} =$	$\frac{7}{10}$ x 30 =
$\frac{7}{9} + \frac{5}{18} =$	$7 - 2\frac{2}{5} =$	$\frac{2}{3}$ x 18 =
$3 - \frac{7}{12} =$	$5 - 4\frac{3}{5} =$	$\frac{3}{4}$ x 32 =
$6 - \frac{2}{5} =$	$10 - 3\frac{1}{10} =$	$\frac{4}{5}$ x 25 =
$\frac{9}{11} + \frac{9}{22} =$	$12 - 11\frac{4}{9} =$	$\frac{5}{7}$ x 14 =
$5 - \frac{7}{8} =$	$15 - 10\frac{89}{100} =$	$\frac{3}{11}$ x 66 =
$\frac{2}{3} + \frac{11}{12} =$	$\frac{1}{6} +$ ______ $= 4$	$\frac{3}{8}$ x 56 =
$\frac{13}{24} + \frac{5}{8} =$	$2\frac{1}{6} +$ ______ $= 4$	$\frac{2}{3}$ x 30 =
$\frac{5}{7} + \frac{17}{21} =$	$7\frac{4}{5} +$ ______ $= 10$	$\frac{5}{12}$ x 24 =
$4 - \frac{5}{18} =$	$3\frac{4}{9} +$ ______ $= 8$	$\frac{3}{5}$ x 100 =
$12 - \frac{7}{9} =$	______ $+ 5\frac{5}{6} = 7$	$\frac{3}{4}$ x 100 =

1 0 0 0	1 0
2 0 0 0	2 0
3 0 0 0	3 0
4 0 0 0	4 0
5 0 0 0	5 0
1 0 0 0 0	1

6000	60
7000	70
8000	80
9000	90
100	200
20000	2

300	400
500	600
700	800
900	000

30000	3
40000	4

50000	5
60000	6
70000	7
80000	8
90000	9
0000	00

100000

200000

300000

400000

500000

600000

700000

800000

900000

00000

0

5000	6000	7000				
					20,000	
29,500	29,600	29,700				30,100
			28,800			
24,230						
24,130			26,800		60,000	
24,030					70,000	
			24,800			
23,830	23,820	23,810				23,770
23,630		23,650		23,670		23,690

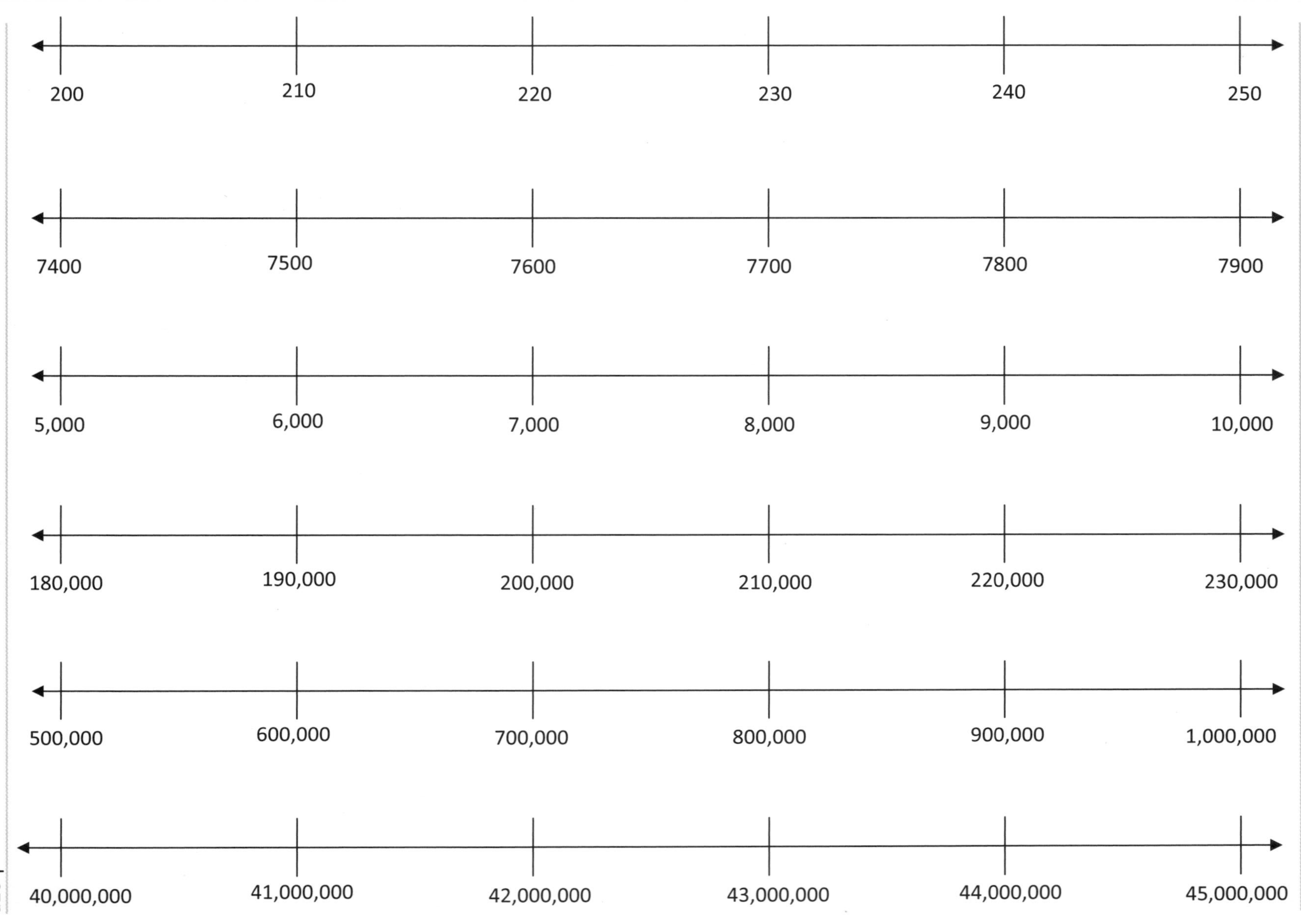
200
210
220
230
240
250
7400
7500
7600
7700
7800
7900
5,000
6,000
7,000
8,000
9,000
10,000
180,000
190,000
200,000
210,000
220,000
230,000
500,000
600,000
700,000
800,000
900,000
1,000,000
40,000,000
41,000,000
42,000,000
43,000,000
44,000,000
45,000,000

4	2	8	5
6	3	☆	12
11	6	7	13
9	2	10	3
4	5	7	☆

1	2	3	4	5	6	7	8	9	10
11	12	13	14	15	16	17	18	19	20
21	22	23	24	25	26	27	28	29	30
31	32	33	34	35	36	37	38	39	40
41	42	43	44	45	46	47	48	49	50
51	52	53	54	55	56	57	58	59	60
61	62	63	64	65	66	67	68	69	70
71	72	73	74	75	76	77	78	79	80
81	82	83	84	85	86	87	88	89	90
91	92	93	94	95	96	97	98	99	100

7 + 10

7 + 9

37 + 9

637 + 9

3637 + 9

637 + 100

637 + 99

3637 + 99

3637 + 1000

3637 + 999

1358 + 7

1358 + 97

1358 + 997

42 – 10

42 – 9

342 – 9

7342 – 9

342 – 100

342 – 99

7342 – 99

7342 – 1000

7342 – 999

2345 – 7

2345 – 97

2345 – 997

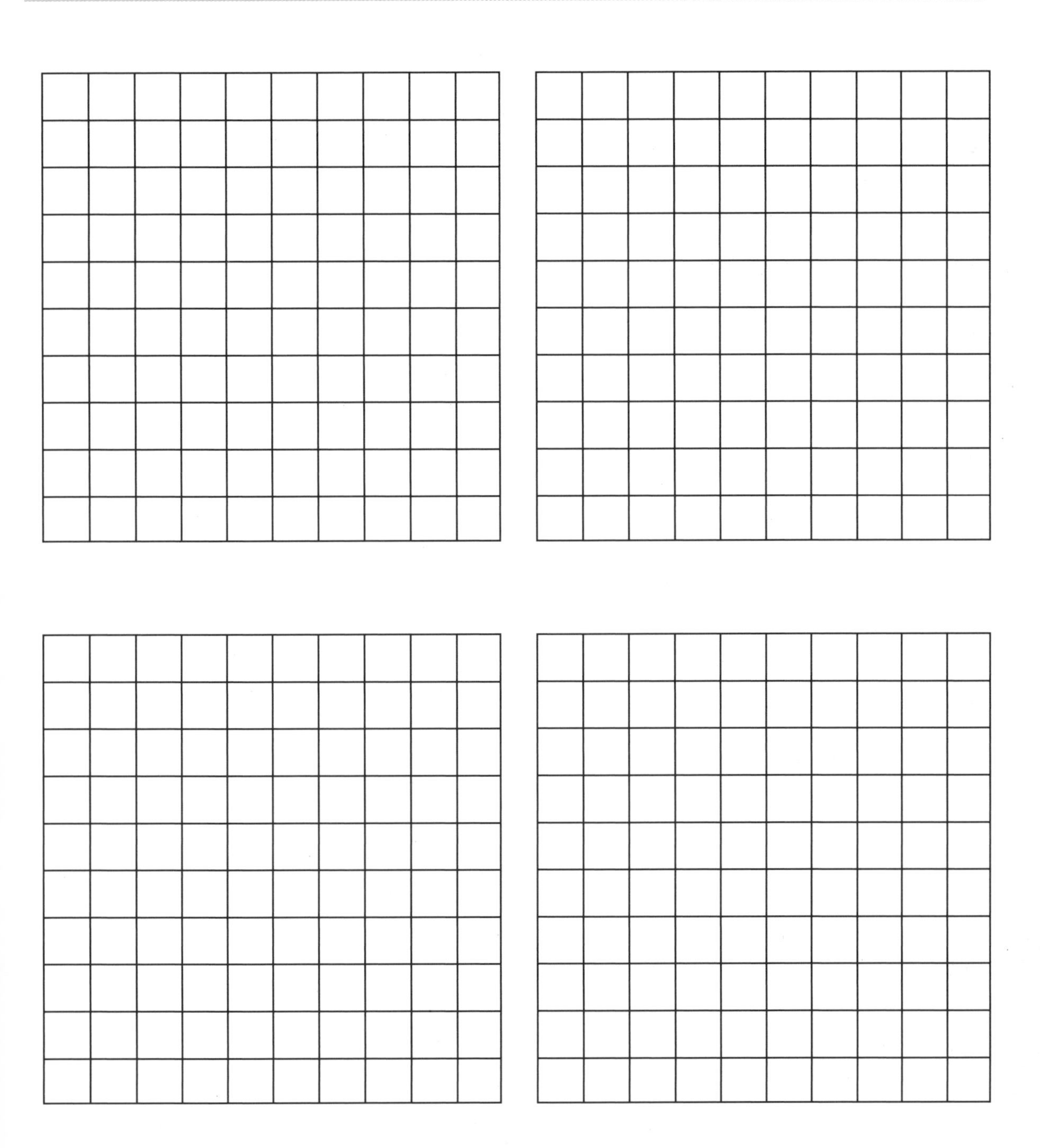

Identify perpendicular lines

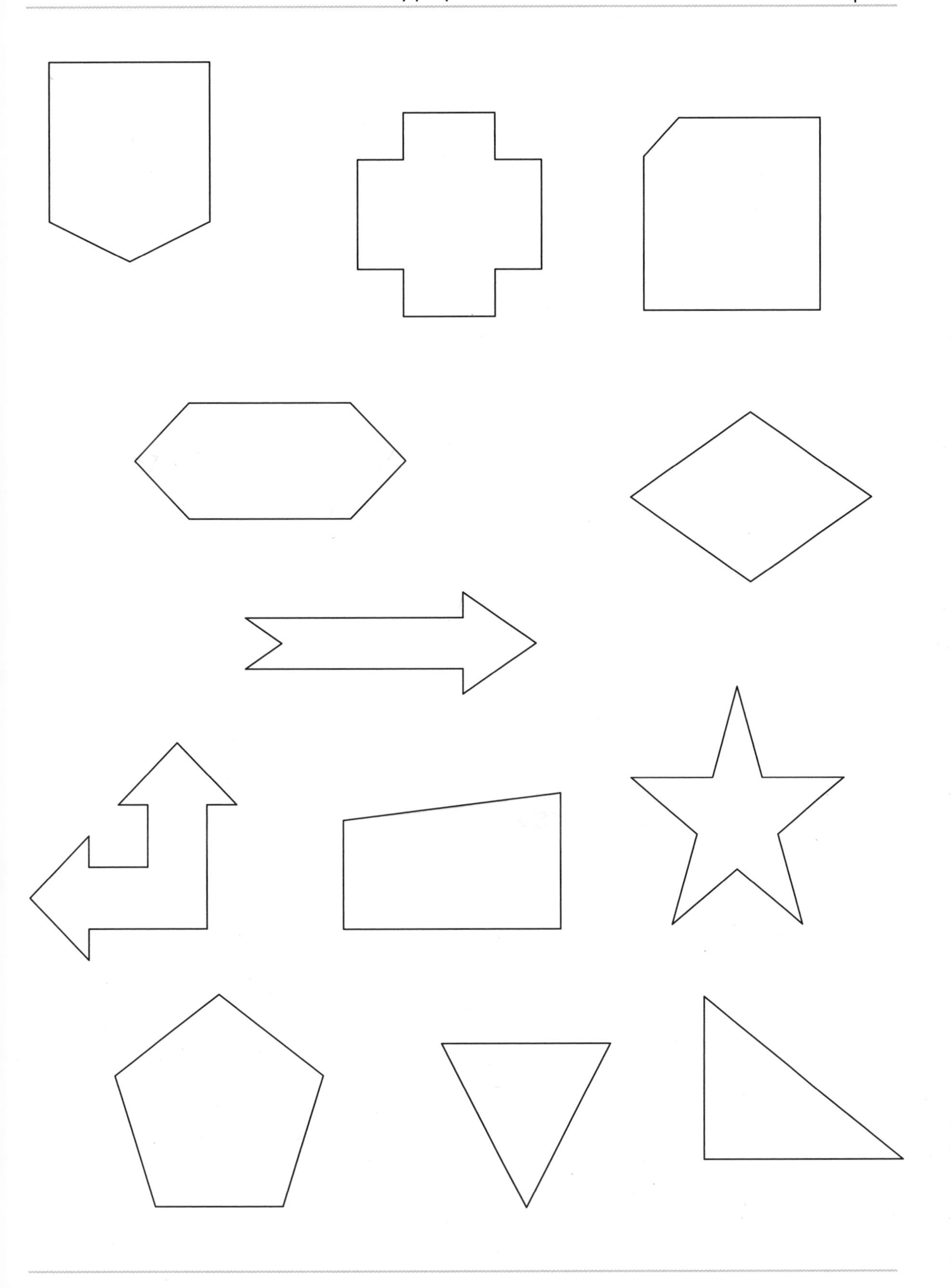

Identify parallel lines